Disclaimer

The publisher of this book is by no way associated with the National Institute of Standards and Technology (NIST). The NIST did not publish this book. It was published by 50 page publications under the public domain license.

50 Page Publications.

Book Title: Chemical Science and Technology Laboratory (CSTL) 2009 Annual Report

Book Author: Willie E. May; Richard R. Cavanagh; Dianne L. Poster; Michael D. Amos

Book Abstract: CSTL is entrusted with building, sustaining, and maximizing the chemical measurement system that is criticial to chemical technological innovation, economic competitiveness and new job growth for the benefit of the Nation.

Citation: NIST Interagency/Internal Report (NISTIR)

Chemical Science and Technology Laboratory

2009 Annual Report

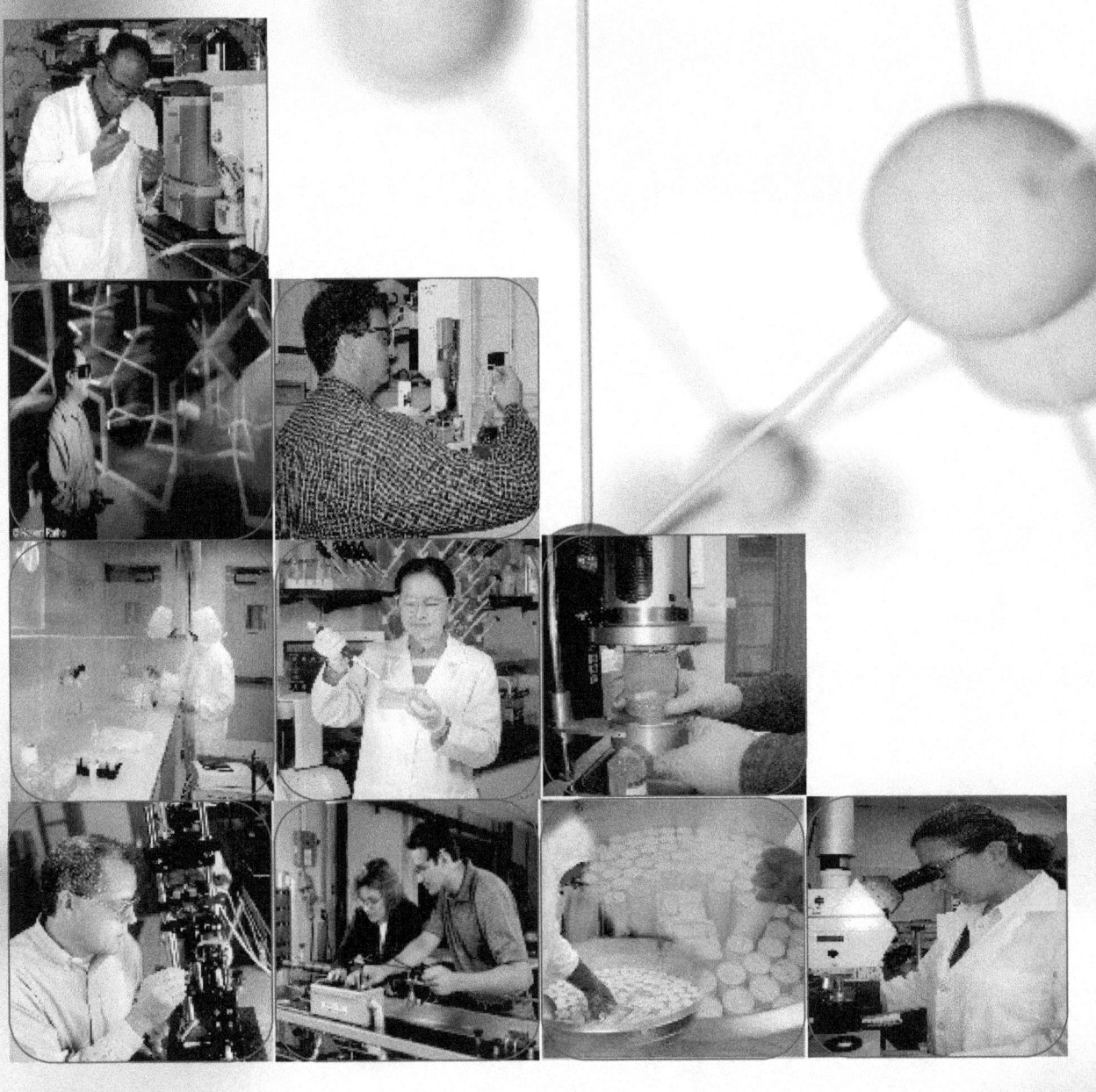

What We Do and Why You Should Care

FACTS & STATS	4-5
LETTER FROM THE DIRECTOR	6-14
STRUCTURE & PROPERTIES	15-22
WELL & WISE	23-42
FUEL & FEED	43-54
SAFE & SECURE	55-64
ELECTRONS & VOICE	65-70
SMALLER & SMALLER	71-82
MEASUREMENTS & STANDARDS	83-92
YEAR IN REVIEW	93-108
FINANCE & INVESTMENTS	109-112

CSTL is entrusted with building, sustaining, and maximizing the chemical measurement system that is critical to chemical technological innovation, economic competitiveness and new job growth for the benefit of the Nation.

IMPROVE the QUALITY of LIFE

The Chemical Science and Technology Laboratory enables the well-being of our citizens every day by its measurement and standards work. At a minimum, the quality of the water we drink, the air we breathe, and the food we eat depends in part on that work. By addressing societal needs for measurements, standards, data, and assessment of new technologies in the areas broadly encompassed by chemistry, the biochemical sciences, and chemical engineering, the Laboratory fulfills the mission of the U.S. Department of Commerce National Institute for Standards and Technology, which reads: *To promote U.S. innovation and industrial competitiveness by advancing measurement science, standards, and technology in ways that en-hance economic security and improve our quality of life.*

Facts & Stats
CSTL At-A-Glance

2 Presidential Early Career Award for Scientists and Engineers winners are among current CSTL researchers.

325 CSTL employees in Gaithersburg, Maryland, Boulder, Colorado, and Charleston, South Carolina support the mission of CSTL.

89 patents have resulted from CSTL research since 1975, with an additional 8 currently pending.

1,000 calibrations and tests performed each year by CSTL for customers worldwide.

Chemical Science and Technology Laboratory — 2009 Annual Report

31 countries represented in CSTL by guest researchers in 2009 alone.

19 CSTL postdoctoral fellows with expertise aligned with CSTL priorities have been retained as full time staff since 2003.

1 the first SRM issued by the National Bureau of Standards in 1901 is still produced by CSTL today as SRM 1d, Argillaceous Limestone.

2,500 websites link to the CSTL developed Chemistry WebBook including essentially every technical library in the world.

26,500 chemical standards produced by CSTL constitute the 32,000 NIST SRM units sold annually.

2,000,000 miles of pipe CSTL contributes to in measuring critical flow for natural gas.

LETTER FROM THE DIRECTOR

Dear Stakeholders:

All of us in the NIST Chemical Science and Technology Laboratory (CSTL) are proud to share with you our 2007 - 2008 Annual Report. The Chemical Science and Technology Laboratory is one of ten discipline-based technical operating units at the U.S. Department of Commerce's National Institute of Standards and Technology and serves as the Nation's primary reference laboratory for chemical measurements, standards, and data, to promote commerce, improve quality of life, and innovation in the areas broadly encompassed by chemistry, the biosciences, and chemical engineering.

CSTL core competencies are:

- Providing traceability to SI or other internationally recognized stated references for:
    - Identification and quantification of chemical and biochemical species/parameters
    - Characterization of the spatial and temporal distribution of chemical and biochemical species
    - Fluid quantities, humidity, pressure/vacuum, leak rate, thermometry, liquid density, volume, air speed, pH, and electrolytic conductivity
- Providing reference chemical, biochemical, thermophysical and kinetic property data and information for chemical, biochemical, chemical engineering and related technologies.

CSTL core competencies are underpinned by a world-class research program in measurement science that enables CSTL to address current and future standards and data needs and to underpin the development, implementation and/or assessment of new technologies. We establish and maintain scientifically sound, metrologically-based competencies and measurement capabilities that are internationally vetted and recognized. With this infrastructure, we provide calibration and measurement services that are disseminated to our customers via various mechanisms such as validated reference methods, certified reference materials, reference data, measurement services for other government agencies, etc. This research basis coupled with extensive "conversations" with our customers and with other national metrological organizations enables us to continue to adapt our chemical measurement services portfolio and provide world class services that are relevant and fit-for-purpose.

CSTL's research and measurement service programs are fueled by more than 350 chemists, physicists, engineers, and biologists at our two primary campuses (Gaithersburg, Maryland and Boulder, Colorado) and at the Center for Advanced Research in Biotechnology in Rockville, Maryland and the Hollings Marine Laboratory in Charleston,

South Carolina. These programs are carried out in six divisions: Analytical Chemistry, Biochemical Science, Chemical and Biochemical Reference Data, Process Measurements, Surface and Microanalysis Science, and Thermophysical Properties. Our technical excellence, broad range of capabilities, and discipline-based organizational structure provide the flexibility to respond to the ever-changing needs of the nation.

Our national economic prosperity and security require that we remain a world leader in science and technology. Education and training for American students is the foundation of a secure U.S. technology leadership. Over the past two years, CSTL has provided hands-on research experiences for over 25 high school students. During that same period, ~120 Undergraduate Students and ~40 Postdoctoral Researchers have conducted research in CSTL laboratories.

The research and measurement services that we provide are critical to virtually all industrial sectors and technology areas. Our customer base is extensive, ranging from established industrial sectors and emerging industries to government agencies, standards and trade organizations, and the academic and scientific communities. The following are a few examples of measurement services provided to our customers in the recent past:

- At the request of the National Institutes of Health Office of Dietary Supplements, the Analytical Chemistry Division has developed a Certified Reference Material (SRM 3280 Multivitamin/Multielement Tablets) to serve as an accuracy benchmark for analytical methods to be used for acquiring data for a Dietary Supplement Ingredient Database being developed by NIH-ODS and the U.S. Department of Agriculture to assess the nutritional intake of the U.S. population. SRM 3280 has concentration values assigned for 24 elements and 17 vitamins/carotenoid compounds.

- The Chemical and Biochemical Reference Data Division has been working with the National Institutes of Health and the National Cancer Institute in developing quality control metrics and serving as an QA/QC advisor for a major inter-laboratory program examining variability in mass spectrometry-based proteomics experiments (the "Clinical Proteomic Technology Assessment for Cancer" program). More recently, scientists in the Division have been working in expanding research programs leading to the analysis of complex, practical reference standards containing thousands of components observed by LC-MS in biological samples. This will involve studies of reproducibility, development of data analysis methods, and creation of reference spectral libraries for use with the reference standards.

- The Thermophysical Properties Division has recently completed a suite of studies on ten fuels, based on needs expressed by stakeholders at a series of workshops. Experimental property measurements, studies using an advanced distillation curve method, and stability studies, were followed by the development of models for the fuels' behavior over extensive ranges of temperature and pressure. This program largely dealt with the military's need for new fuels and fuel standards for the next generation of re-usable rockets, but feeds into a broader program on renewable fuels.

- Staff from our Surface and Microanalysis Science Division have completed a seminal study of the trophic transfer of nanoparticles in a simplified invertebrate food web in response to the research needs outlined by the U.S. Government's National Nanotechnology Initiative in its document "Environmental, Health, and Safety Research Needs for Engineered Nanoscale Materials." This study addressed the potential environmental risks of engineered nanomaterials on aquatic organisms by demonstrating that nanoparticles can be transferred from one aquatic organism to a higher trophic level organism. Although limited bioconcentration and lack of biomagnification may impede the detection of nanomaterials in invertebrate species, the findings of this study indicate that dietary uptake should be considered as a potential source of nanomaterials for higher trophic quatic organisms.

- The Biochemical Science Division has a complementary program in the area of nanotoxicology. This program involves the testing and validation of existing in vitro toxicological screening assays that are applied to measure the toxicity of engineered nanomaterials; as well as the development of new in vitro assays with the goal of understanding the mechanism of nanomaterial toxicity. To complement our work in cell-based mechanistic assays, we are using metabolomics and proteomics to evaluate the impact of engineered nanomaterials at the molecular level. Additionally, we study genotoxic effects of these materials leveraging our well established program in DNA damage and repair. Collaboration across NIST with nanomaterials researchers in CSTL's Surface and Microanalysis Division and NIST's Materials Science and Engineering and Physics Laboratories is critical to the success of this program. Equally important is working with our federal partners that include the FDA, EPA and NIH who have established programs in nanotoxicology as part of their mission. A broader understanding of the needs of this community is gained through many external interactions including participation on the ISO Nanotechnologies Technical Committee.

- The Biochemical Science Division's Human Identity Team has spent the past several years building a national infrastructure in partnership with the Department of Justice to promote, ensure, and enable excellence in DNA analysis for human identity. This national infrastructure includes validated, advanced DNA measurement methods; SRMs to calibrate instrumentation and validate measurement results; methods for data evaluation and interpretation; SRDs; teaching tools; and the definitive textbook (2nd Edition) on forensic DNA analysis. These tools are relied upon to assure accurate assessment of human identity based on DNA obtained from crime scenes or mass disasters (e.g., Hurricane Katrina, the WTC disaster), military operations, and criminal investigations).

- The Process Measurements Division has established a calibration service for flow meters for natural gas at pipeline conditions. Starting with NIST's on-site gas flow standards, the Division achieved a 600-

fold increase in flow-rate measurement capability at a 10-fold increase in operating pressures, thereby enabling the calibration of large (up to 30 inch diameter) meters used to accurately measure the flow of natural gas as it is imported in pipelines. (Imports are $20B/year.) This calibration service also satisfies the European Union's requirement that all natural gas flow meters be calibrated before entering commercial service. Thus, the service enables US manufacturers and calibrators of flow meter to compete more effectively internationally.

The World is changing and NIST and CSTL must respond accordingly. NIST's strategic priority areas are: National Physical Infrastructure, Renewable Energy, the Environment (including Climate Change), Health Care, Secure IT Systems and Communications, and Homeland Security. Initially determined based on discussions at a 2006 CSTL Strategic Planning Retreat and reinforced by alignment with recently articulated NIST Priorities, CSTL's areas of strategic emphasis are measurement science, standards and data to support national efforts and interests in:

- Bioscience and Health
- Climate Change Assessment
- Renewable Energy
- Nanotechnology

In the area of bioscience and health, we are focusing on standards for clinical diagnostics and genetic testing, cellular and molecular imaging, and biopharmaceutical manufacturing. In the area of climate change, we are focusing our efforts on standards for greenhouse gases, investigating effects of aerosols on radiative forcing, thermophysical properties to support determination of global warming potentials of atmospheric species, and measurement and standards to support a market-based national "Carbon Cap and Trade" program. With respect to renewable energy, we are developing certified reference materials for bioethanol and biodiesel from various sources and several projects that will support the equitable trade of hydrogen fuel. In the area of nanotechnology, new efforts are focused on investigations of the environmental, health and safety effects associated with use of engineered nanomaterials and the measurement science to enable spatially-resolved chemical analysis at the nanoscale.

As you will see in this report, CSTL's research and measurement service programs impact all parts of our society. For example:

- **The $700 billion international chemical process industry relies on quality data in order to efficiently sustain its infrastructure with flexible feedstocks, changing product demands, and innovative processes. CSTL currently has more than 30 data projects, striving to provide a variety of industries with reliable information including robust uncertainty estimates. The portfolio of CSTL data ranges from an HIV Structural Database for AIDS Research to Gas Phase Ion Thermochemistry. CSTL data products currently are used by more than 70 000 corporate clients, and are in essentially all mass spectrometers delivered to the research and laboratory communities. CSTL is a major player in helping deal with the abundance of information, through standards work dealing with data exchange and automated systems for data evaluation.**

- **CSTL is responsible for over 1000 Standard Reference Materials (SRMs) that are used worldwide to calibrate chemical measurement instrumentation and/or validate the accuracy of measurements ranging from spectrophotometric filters and calibration solutions to natural-matrix materials for clinical, food, industrial (cement, fuel), and environmental applications, to forensic DNA materials and materials for surface microanalysis. CSTL** strives to provide these communities with the tools they need to fulfill their missions. In FY08, NIST sold more than 27,000 SRM units that originated in CSTL, generating $1.8M in sales. SRMs originating in CSTL are currently used by more than 3000 unique customers.

- Each year, CSTL calibrates approximately 500 instruments that measure temperature, fluid flow, pressure, vacuum, humidity, volume, and liquid density. Most of these calibrations are at the highest, i.e., "best-in-the-world," level. Calibrations provide traceability from U.S. national measurement standards to customers that include instrument manufacturers, military and private calibration laboratories, and other Federal and State agencies. Instrument calibrations enable our customers to verify the reliability of their in-house standards and calibration procedures. Moreover, many commercial customers require traceability to NIST to sell their products and calibration services in international markets that demand traceability to National Metrology Institutes such as NIST.

In this report, we summarize selective elements of these activities in an effort to illustrate the wide breadth of scientific achievements attained by CSTL scientists and to call attention to the multiplicity of measurement services that we provide to our Nation and the world. CSTL's measurement science and standards form an integral part of the foundation upon which innovation is built. This report represents only a snapshot of our activities; the most complete picture is available on our web site: www.nist.gov/cstl.

In closing, I want to thank our dedicated staff, who are committed to ensuring that CSTL continues as a dynamic, flexible, innovative, and productive entity. As no one person or organization can address all the Nation's measurement, standards, and technology needs, it is important to recognize the contributions of our many collaborators and partners in industry, government, standards and trade organizations, as well as the academic and scientific communities. To be most effective and successful, we need their ongoing input and partnerships. We are proud of the work that we have accomplished and feel that given the broad demand for our services, CSTL has performed admirably in helping to fulfill the NIST mission, but we realize that a lot more will be required for meeting the enormous challenges of the 21st Century.

Willie E. May, Ph.D.

Director, Chemical Science and Technology Laboratory
National Institute of Standards and Technology
Gaithersburg, MD 20899 301-975-8300

cstlinfo@nist.gov   www.nist.gov/cstl

Structure & Properties
Chemistry Research Activities

Chemistry is a branch of science that seeks knowledge on the chemical composition, structure, or properties of a substance, all to the tune of $664 billion a year, the value of the American chemical industry enterprise. CSTL chemistry research activities provide critical reference data, data-predictive computational tools and instrument techniques for this enterprise. These products make available flexible, error-tolerant, world-recognized information and tactics necessary for analysis, control, and optimization of processes pertinent to the chemical technology sector.

A PORTAL TO CHEMICAL PROPERTIES

Working with international standards body has enabled the dissemination and interchange of data for chemical systems worldwide.

Imagine a website with more than sixty thousand page views per day and you have a sense of the CSTL developed NIST Chemistry WebBook where anyone can search for data on specific compounds based on a chemical's properties. Now imagine integration of databases from multiple sources, including many developed by NIST, to continuously build up and enhance the utility of WebBook, today such a resource does not exist. That is why CSTL researchers, working with the International Union of Pure and Applied Chemistry have developed the International Chemical Identifier for the identification of chemical species to overcome hurdles with the integration of new data into the WebBook. This tool makes it easier for chemical species (and mixtures and reactions) to be reliably identified and offers the most accurate and up-to-date information for chemical systems to meet high priority industrial process and design research needs. The approach also enables standardized referencing and searching for chemical structures over the Internet and in proprietary databases.

Contact
Peter J. Linstrom
Dmitrii V. Tchekhovskoi

VIRTUAL MEASUREMENTS INCREASE THE VALUE OF PREDICTED PROPERTIES

Predictions from quantum chemistry with meaningful uncertainties will enable drop-in replacements for costly experimental measurements.

By itself, a predicted value for a quantity constitutes merely a guess. To be a proper measurement, the associated uncertainty must be supplied. Most values reported from ab initio calculations, an abbreviation referring to quantum chemistry methods, are not accompanied by uncertainties. As a result, the reliability of such data, such as reactive-chemical hazards, is poor. Safety thus requires skepticism, often leading to expensive over-design. CSTL researchers are quantifying the uncertainties associated with ab initio predictions, thus putting these virtual measurements on an equal footing with experimental measurements and increasing their dollar value. Predictions related to applications most relevant to chemical engineering, analytical chemistry, and physical chemistry, are under investigation in collaboration with the NIST Information Technology Laboratory and are part the NIST Computational Chemistry Comparison and Benchmark Database on the Internet. These results allow industries to incorporate predictions in their R&D processes with a degree of quantitative reliability.

Contact
Russell D. Johnson III
Karl K. Irikura

Structure & Properties

A NEW GENERATION OF CHEMICAL ANALYSIS

Novel high-resolution x-ray spectroscopy identifies chemical compounds that may be explosive or toxic.

X-ray emission spectra, in combination with the latest computational theoretical chemistry, provide insight to the bond structure of chemical substances. Knowledge of the bond structure is needed to understand the chemical information in x-ray data. CSTL researchers, with collaborators from the University of Washington and Lawrence Berkeley Laboratory, are investigating the basis of x-ray spectra on energetic nitrogen compounds to further the development of new high-resolution x-ray spectrometers for chemical analysis. These nitrated compounds, which are commonly used for explosives in military and industrial applications, are important to study because they exhibit unusual chemical behavior, incorporate multiple nitrogen atoms in inequivalent sites, and have sophisticated chemical structures. It is possible that x-ray emission spectroscopy will be useful in the future detection and identification of these compounds.

Contact
Terrence J. Jach

UNDERSTANDING MOLECULES WITH MANY THOUSANDS OF ATOMS

A fast and efficient quantum mechanical modeling method coupled with high performance computing and visualization accelerates scientific discovery.

Many current chemical problems involve huge numbers of atoms, in fact many thousands of atoms. Examples include the simulation of proteins in aqueous solution to study new drug therapies and prediction of the behavior of nanoscale electronics to advance semiconductor and data-storage research. In collaboration with scientists from Venezuela, CSTL researchers have created innovative quantum chemical techniques, specifically tight binding methods based on density functional theory, to permit calculations of properties and simulations of dynamic chemical processes for molecules with many thousands of atoms. This capability facilitates the study of such compounds in realistic chemical environments. Ultimately, these methods can aid in the design of molecules targeting specific properties. Moreover, the use of these methods overcomes severe computational limits associated with large molecular systems and enables exploration on resources, such as a desk top computer, that are readily available to researchers.

Contact
Thomas C. Allison

SOLVING NUMEROUS MYSTERIES OF DETONATION

Understanding the fundamentals of explosive detonation involves serious number crunching.

Detonation, an event that lasts less than a millionth of a second, involves a huge range of interplay among the chemical and physical properties for each of the compounds in the device. Very rapid chemical reactions occur with the evolution of large volumes of heated gases that exert high pressures in the surrounding medium. The challenge for scientists who study explosive detonation, as well as deflagration, combustion, and slow ageing, is to identify these reactions over a wide range of pressures, densities, and temperatures and the way these interact. Working with the U.S. Army and Air Force research laboratories, CSTL scientists are using high-level quantum chemistry to verify the most important processes during detonation of high-energy materials and are including combustion modeling as a means to better identify the most important gas-phase reactions. The findings will benefit new explosives design, new strategies for detecting explosives and the engineering of delivered systems such as military ordnance.

Contact
Karl K. Irikura

EXPLOITATION OF FLUID PROPERTIES GUARANTEES ADVANCING SMALL-SCALE TECHNOLOGIES

Fluids confined in small spaces are ubiquitous in the natural and industrial world, and are of great interest for their practical importance in new technologies.

The properties of confined fluids, such as temperature, volume, and pressure, in channels engineered on a scale not even visible under a microscope are a current research area for nanotechnologies. Properties of confined fluids are critical to designing substrates for technological applications (e.g., membrane separations in chemical processes and materials synthesis, chromatography, fluid capture, storage and release, and heterogeneous catalysis). Unfortunately, the properties of a fluid can change profoundly when placed in a confined environment and currently there are no theoretical tools for predicting these changes. As a result, advancing the use of fluids in technologies that operate at very small scales is difficult at best. In an effort to better enable the development of nanotechnologies that make use of liquid phenomena, CSTL researchers are conducting comprehensive computational studies of simple fluids in nanoscale confinement. Many of the simulations incorporate specialized computer codes designed to answer specific questions posed by different phenomena. The findings have provided a trouble-free procedure for predicting the behavior of fluids in confined spaces over an enormous range of liquid states spanning from a dilute gas to a supercooled liquid (where the liquid exists below its freezing point without becoming a solid). This modeling capability makes possible design, development, and optimization of liquid processes at the nanoscale for medicine, food production and packaging, chemical pigment production and fuels.

Contact
Vincent K. Shen

Structure & Properties

BRIDGING THE GAP: SOLVING THE MICRO-TO-MACRO PROBLEM

A novel technique has provided direct evidence of the penetration of potassium ions from deicing salts as deep as 10 cm from the surface of airport runways.

For much of our Nations' physical infrastructure, deterioration represents a major cost, both in terms of maintenance and replacement. Concrete, a pervasive component of the Nations 'transportation infrastructure, is such an example. Deterioration of airport runways in areas where deicing solutions have been applied has recently received attention. While an extensive suite of analysis tools currently provides characterization of the chemical composition at the microstructural scale, to answer degradation issues, it is essential to be able to extrapolate from the "micro" scale where the existing tools are effective, to the "macro" scale. Chemical metrology tools that allow both a macro scale assessment of the chemical changes associated with materials degradation, and also the chemical changes on a microscopic scale. It is on this latter scale where research in the concrete industry strives to improve performance. For those problems, we need to take a broader view of the chemical changes, out to several centimeters. We have explored an important alternative based on a commercial x-ray fluorescence instrument. The basic instrument mapping function has been operated for the first time in an "x-ray spectrum imaging" mode and applied to concrete that had been exposed to deicing solutions. The approach was capable of distinguishing native potassium in the pre-existing mineral assemblages of the poured concrete, from the intrusive deicing potassium that severely weakened the concrete. The method will be widely used by the chemical, chemical engineering, and material science industries for better and safer product development.

Contact
Jeffrey M. Davis

Well & Wise
Bioscience and Health Research Activities

Biology is an informational science that critically depends on accurate measurements and standards. Whether quantifying the amount of protein in a cancer cell or how well an organism converts sugar to alcohol, measurements are the foundation for improving our understanding of biological systems to meet the next generation of health care needs. What's next depends on our ability to measure key features— including the complex interplay of thousands of biochemicals that control living systems. CSTL activities are focused to meet these challenges by advancing protein measurements, bioimaging, DNA diagnostics, and health care measurements.

REACHING TOWARD A VISION OF THE FUTURE IN BIOMANUFACTURING

Better, faster, safer protein drug manufacturing requires new approaches to measure proteins.

On-line process tools, improved sensor calibration and analytical and data analysis methods are among the technological challenges facing the biomanufacturing industry over the next 10 years. These challenges must be addressed in order to greatly increase manufacturing efficiency for protein drugs while ensuring safety. Improved measurement standards and technologies are essential initial steps towards the "blue sky vision" for the future of biomanufacturing. CSTL scientists took a step toward this vision by developing electrospray-differential mobility methods to measure the size and three-dimensional shape of aggregates of therapeutic proteins, one of the fastest growing pharmaceutical markets. This new technology provides for safer protein drugs through better strategies to limit or prevent aggregation, which can be catastrophic. In 1998, an undetected aggregation of a protein drug caused patients to develop an immune response to the aggregates that destroyed their ability to make their own red blood cells. Many of these patients will likely require blood transfusions for the rest of their lives. These research findings could prevent such devastating events.

Contact
Michael J. Tarlov
Michael R. Zachariah

Chemical Science and Technology Laboratory
2009 Annual Report

SHATTERING PROTEINS TO REVEAL CODES OF LIFE

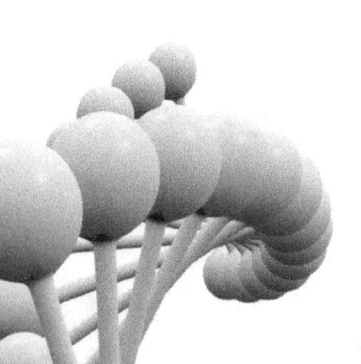

Reliable, routine identification of peptides will lead to discovery of new biomarkers and clinically successful diagnostics based upon analytical proteomics.

Proteins are polymers of amino acids responsible for the work and health of the body. Proteomics research aims to understand these critical proteins, their content and their function. Such research will lead to better diagnoses of disease based upon associated changes in proteins makeup and concentrations in human cells and fluids. CSTL scientists are helping to enable the use tandem mass spectrometry to determine the sequence of amino acids in proteins for better disease diagnostic tools. This technique breaks large proteins into small peptide pieces and measures the masses of these smaller pieces. The fragmentation pattern can be used as a fingerprint for identifying a particular protein within a mixture of other proteins, a requirement for disease diagnostics. However, many mass-spectral peaks remain unidentified and only partial fingerprints are available. Using quantum chemistry to explain known trends in peptide fragmentation and to predict reaction mechanisms in the tandem mass spectrometry environment, CSTL researchers are developing fragmentation rules to accurately predict mass spectra of peptide fragments and increase the power of peptide analysis in proteomics research.

Contact
Karl K. Irikura

DESIGNING STABLE PROTEINS, EVEN IN CROWDS, FOR SUPERIOR PHARMA

Knowing if a protein favors a biologically active state or an inactive, denatured form can mean life or death.

Our knowledge of protein folding, the process by which a polypeptide folds into its function-essential, three-dimensional structure, comes from experiments in dilute solutions or from theoretical models of isolated proteins. However, neither biological cells nor solutions in protein-based drug development, are dilute. Rather, they are "crowded" with proteins, sugars, salts, DNA, and fatty acids. Crowds can effect protein unfolding, aggregation, and precipitation, all of which may lead to loss of protein functions. Laboratory studies of crowd effects are expensive and time-consuming but computational methods offer relief. In collaboration with scientists from the Schering-Plough Research Institute, the State University of New York at Buffalo, and the University of Texas at Austin, CSTL researchers have developed models rich enough to capture critical folding problems, such as the intrinsic free energy of folding in solvent, and allow for the simulation of hundreds to thousands of foldable protein molecules in solution. The end result is a framework that allows one to independently vary, and isolate for detailed study, the parameters most important for effective and safe biotherapeutic drugs, even in crowds.

Contact
Vincent K. Shen

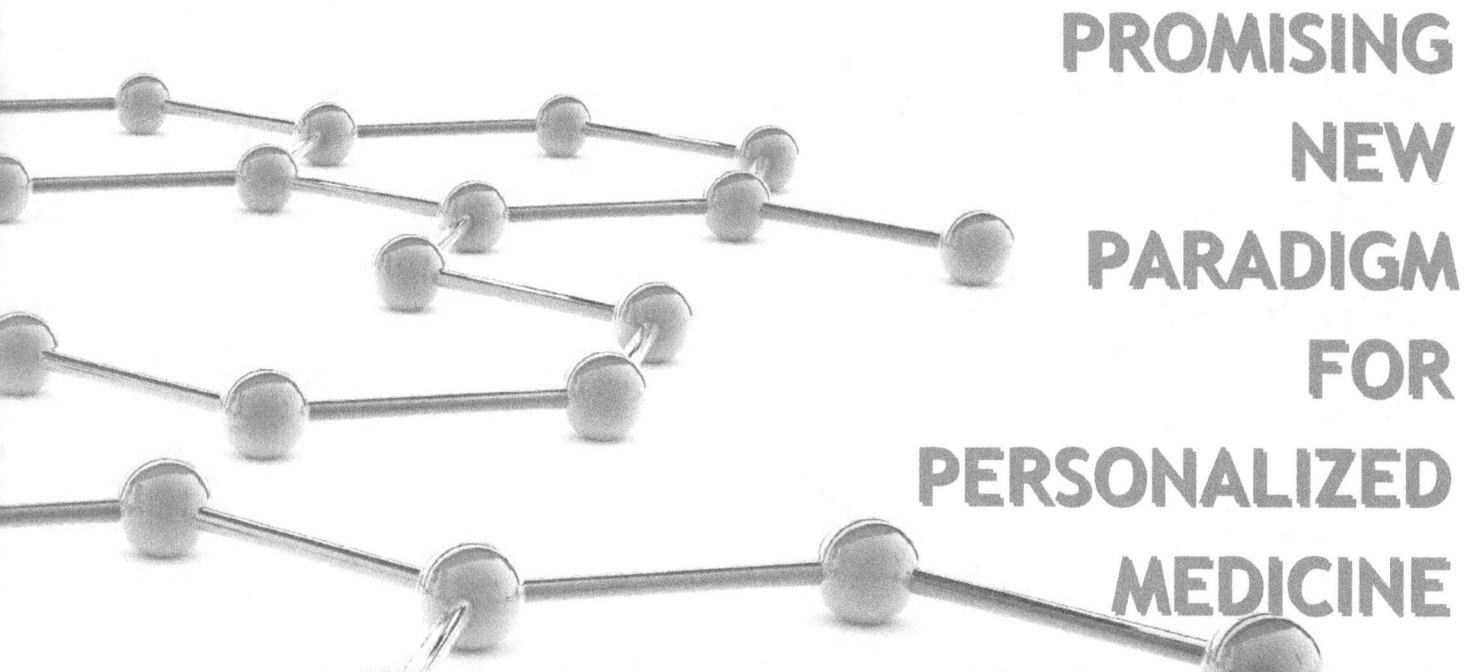

PROMISING NEW PARADIGM FOR PERSONALIZED MEDICINE

Achieving accurate measurements of proteins in blood is critical for advancing proactive, preventive medicine.

Serum proteomics, the measurement of the complete collection of proteins associated with health and diseases is a key element to meeting the challenges to diagnose early stages of disease via the identification of altered patterns in the proteome. However, several problems in serum proteome measurement sensitivity and reproducibility remain unresolved. Standard Reference Materials (SRM) help address issues such as these and assist laboratories in validating measurements of physical or chemical composition. Currently, a material to assess the quality and comparability of protein measurements in blood is not available. CSTL researchers are filling this gap by developing an SRM that will consist of human serum albumin (HSA), the major protein constituent in blood. As obtaining HSA from human blood supplies is complicated by the possibility of infection with HIV or other agents, CSTL researchers have developed approaches to obtain HSA from the bacterium Escherichia coli using genetic engineering. The harvested HSA will be used for the proteomic SRM, an SRM acutely needed to assess the reliability and reproducibility of current tools used to diagnose and treat disease.

Contact
Prasad T. Reddy
David M. Bunk

CANCER DIAGNOSTICS WITH SNAPSHOT MASS SPECTROMETRY

Imaging chemical changes manifested by controlled expression of selected molecules will lead to improved detection of cancer.

Mass spectrometry-based imaging methods for site-specific localization of biochemical changes within and around single cells, as well as tissue, is paving the way to better methods to determine if cells or tissue are cancerous or healthy. In collaboration with scientists at the Laboratory for Cell Biology within the National Cancer Institute of the National Institutes of Health, CSTL researchers are advancing cluster secondary ion mass spectrometry (SIMS) to image aggregates of *Escherichia coli*, one of the best-studied simple prokaryotic model organisms in biology research. The simplicity of this cell line has enabled the rugged testing of proper cell preparation methods needed for SIMS analysis, and for the verification of the imaging modality to provide biologically relevant chemical signatures. Examination of the chemical signatures of cellular material in cancer cells by utilizing the surface sensitivity and molecular imaging capabilities in SIMs is leading to the ability to identify cancerous cells based upon visual differences from normal cells. This technology is being used to explore site-specific chemical changes of disease progression, thereby leading to clinical diagnostics via identification of disease-specific indicators.

Contact
Christopher Szakal

MIGHTY SMALL DOTS PACK A PUNCH TO BEAT CANCER

Minute dots and tubes are leading preemptive strikes against cancer.

Breast cancers can have multiple copies of the gene HER2, one of a family of genes that regulate growth of human cells. It's important to test breast cancer patients for the HER2 amplified gene or its overexpressed HER2 protein in order for physicians to make informed treatment decisions. Unfortunately, the existing tests for these biomarkers can yield a significant number of false negatives, resulting in patients not receiving the drug (Herceptin) that would cure them, resulting in their death. Very small light-emitting materials called quantum dots, which are about a single nanometer (a billionth of a meter) in size, are showing promise for improving this critical diagnostic test. Researchers from CSTL and the National Cancer Institute are using quantum dots to more reliably detect HER2 biomarkers than existing tests that use conventional fluorescent dyes. The improved sensitivity, about 40-50 percent, is in part because the mighty small dots generate intense colorful light, and stay bright and detectable while conventional fluorescent dyes fade over time. Similarly, CSTL scientists have developed tiny carbon nanotube HER2 complexes to both find and kill breast cancer cells. The research team is extending these applications to other cancers, such as prostate and colon, thereby improving clinical decision making for cancer and establishing the place of tiny materials at the forefront of better cancer therapy.

Contact
Yan Xiao
Miral Dizdar

STUDIES IN CELLS SHED LIGHT ON CELL GROWTH & REPAIR

An understanding of how cells interact with their environment enables scientists to understand healthy and diseased states of the body.

A critical challenge in cell biology is imaging and quantifying the interactions of cells with their extracellular environment, including the active remodeling by cells of their extracellular matrix (ECM) or cell spreading. These observations are necessary to advance cell growth and repair studies pertinent to wound healing, developmental biology, and metastasis of tumor cells. CSTL researchers are looking closely at cell growth and repair activity in the ECM setting using advanced instrumentation such as surface plasmon resonance imaging. This is a sensitive, low-light optical method. The low-light specificity prevents cell damage and enables the cells to thrive, thereby offering long periods of time for live observations. Dynamic changes in cell-substrate interactions such as membrane ruffling have been observed, providing researchers with the opportunity to study the formation of motile cell surfaces that contain a meshwork of newly formed filaments that contribute to healthy cell shape and motion. The technique also permits quantification of protein secretion from cells and surface protein density dynamics, observations pertinent to advancing analytical proteomics.

Contact
Alexander W. Peterson
Anne L. Plant

GETTING THE GENE STORY WITH REPORTER CELL LINES

Fluorescent proteins provide the opportunity to probe the dynamics of gene activity within single cells.

A vibrant light-emitting protein, green fluorescent protein (GFP), has existed for millions of years in the jellyfish, *Aequorea victoria* and gives the species it brilliant light in dark waters. GFP from the jellyfish has long been isolated and used as a research tool in cell biology to study other proteins. Cells can be genetically engineered to produce the protein of interest and then linked to GFP which acts as a "reporter". When produced, the "fusion protein" emits light, which can then be used to monitor its activity in the cell under a microscope. CSTL researchers are using GFP for reporting, or tracing, proteins in cells. Because GFP is continuously degraded by the cell once it's produced, it is important to understand the kinetics of these processes to ensure that the observed changes in cellular signals are not due to the normal variability associated with the production of GFP. The research team is using protein patterning and microscopy to measure GFP degradation rates in individual, live cells, often an intimidating task due to the large numbers of cells that must be examined. The team is overcoming this daunting challenge by developing automation microscopy techniques. As a result, it is now possible to rule out the dependence of the degradation kinetics on GFP intensity. This has provided valuable evidence that GFP is an accurate reporter of gene activity within cells and can be an essential tool for understanding the aspects of functional protein production in the body.

Contact
Michael Halter

CONSISTENT GENE ASSAY PERFORMANCE FOR BETTER DIAGNOSTICS

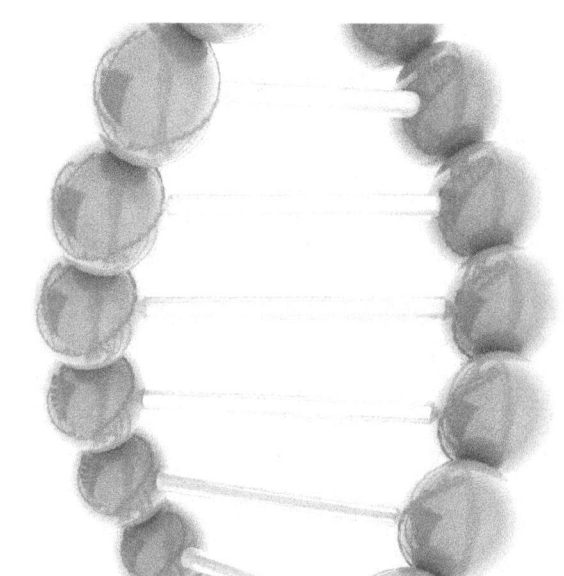

Microarray technology (aka gene chips) is used to determine which fraction of the more than 30,000 human cellular genes is active and which is inactive.

Quantitative measurement and identification of the tiny RNA copies made from DNA when a gene is active using microarray technology is a current application in clinical medicine to predict, for example, the response to chemotherapy of breast cancer patients or the degree of organ rejection in transplant patients. However, variation in performance of microarrays makes it hard to compare results from different manufacturers, or even from a set of tests in the same or different laboratories. Accuracy and reproducibility are extremely important because of the often subtle changes in the many RNAs that must be detected. Even slight uncontrolled variation in the microarrays can cause scientists to miss these subtle changes in the RNA expression patterns, which can lead to poor clinical decisions. CSTL researchers, in collaboration with academic and industry scientists, are devising a way to ensure greater confidence. Reference materials and approaches are being developed that can gauge a microarray's performance during the course of a test. As a founding member and host of the External RNA Controls Consortium, the CSTL team has spearheaded a program to create and promote gene expression microarray standards. The project is entering on a phase of commercial testing, making it possible to compare the performance of different assay designs.

Contact
Marc L. Salit

STANDARDS TO ENSURE WHO'S WHO

Up-to-date reference materials guarantee quality checks for laboratories that make human identifications.

When a crime is committed or a disaster occurs, it is critical to accurately link the DNA found at the scene with the identity of the people from which the genetic information came. This is performed by determining the presence of specific sequences in the DNA that have a very high probability of being from only a single human being. CSTL works with law enforcement and forensics communities to provide materials that forensic laboratories being assessed under the FBI DNA Quality Assurance Standard can comply with. These materials ensure that each laboratory is operating under calibrated conditions, enabling compatible and reliable results among laboratories. CSTL scientists developed standard reference materials based on regions of DNA known as short tandem repeats that can be used to calibrate the measurement tools used to identify humans. Besides the ones used for crime scene investigations, a set of references that are particularly useful for calibrating systems used to analyze degraded DNA, such as that from a fire or explosion, have been developed. In addition, researchers have developed a Standard Reference Material and evaluated computer software that helps forensic laboratories ensure the quality of their work when DNA evidence contains mixtures of material from two or three different people – which is most often the case.

Contact
Margaret C. Kline

AS SEEN ON TV – ON-LOCATION ANALYSIS OF DNA

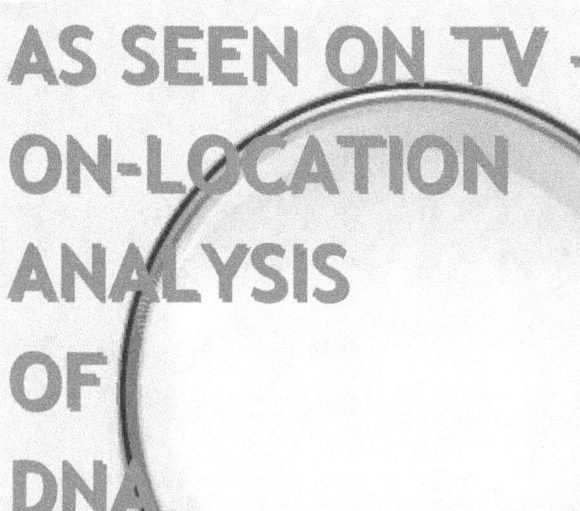

With the continuing development of miniaturization technologies, the overall time to type DNA samples is shrinking too.

When a crime occurs, speed is of the essence when attempting to solve "who done it". Forensic DNA typing is the most accurate method for determining who was present at the scene of the crime. Unfortunately, this type of testing may take several days to complete. Development of miniaturization technologies such as microfluidic and micro-capillary "lab-on-a-chip" devices reduce the overall time required to type DNA samples. Once optimized, these miniature devices could be used for initial screening at a crime scene, at a border, or at airports. CSTL scientists have worked with the forensic laboratory testing community on the development of calibration and quality assurance techniques for the rapid (less than 36 minute) analysis of crime scene DNA. CSTL scientists have established the validity of using Standard Reference Materials in new commercially available test kits for amplification of the minute quantities of DNA left behind at the scene of a crime. Success obtained by the rapid protocols for forensic DNA typing can also be applied to other tests that depend upon DNA amplification, such as clinical DNA diagnostics and DNA biometrics.

Contact
Peter M. Vallone

DETECTING "TECH" IN YOUR FOOD

Trade in grain and foods can be blocked due to inconsistencies in the testing for genetic modifications, leading to financial loss.

The U.S. agricultural system is a major exporter of raw materials such as grain and finished foods. A large part of the market consists of commodity crops. Many of these crops are genetically modified, primarily with traits beneficial to their production. Many importing countries regulate which types of crops are allowed and often require labeling when the concentration of these materials in the food or grain shipment reaches a specified level. Testing is conducted both prior to export and at the point of import of U.S. products. Quantitative real-time polymerase chain reaction is the primary tool for detecting and quantifying the amount of biotech crop material in food, feed or grain. CSTL researchers are addressing the quality of such measurements and the factors which under lie the quality to improve measurements and contribute to international harmonization of testing for biotech crops. The team is also enabling methods to accurately quantify biotech crop material in crops where the DNA is degraded, such as in processed food products. By eliminating inconsistencies and inaccuracies in biotech crop testing, the blockage of trade in grain and foods and the associated financial loss due to such ineptness will be greatly reduced.

Contact
Marcia J. Holden

RELATING A DRUG'S STRUCTURE TO ITS RELEASE IN THE BODY

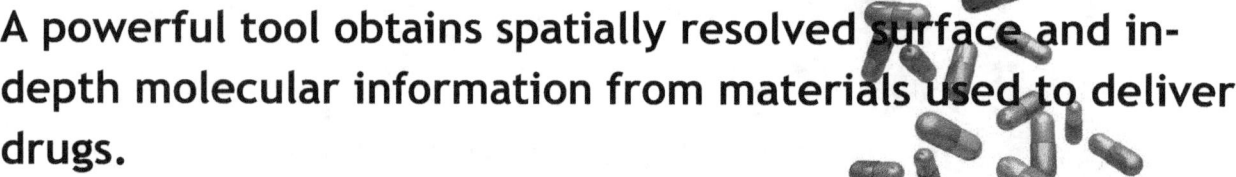

A powerful tool obtains spatially resolved surface and in-depth molecular information from materials used to deliver drugs.

The pharmaceutical industry focuses heavily on analytical techniques necessary to understand interrelationships between drug structure and its processing in the body. Modern techniques to deliver drugs to where they are needed in the body, called drug delivery systems, have complex designs involving multiple layers of polymer materials and intricate compositions. These products and their formulation steps require thorough analyses to determine their functional integrity and reliability and to ensure their effectiveness and safety. CSTL researchers are working closely with the U.S. Food and Drug Administration and industrial partners, such as Medtronic Inc., and Surmodics Inc., to study the release of drugs from various polymer systems using multiple imaging modalities, including laser scanning confocal microscopy, confocal Raman Microscopy, atomic force microscopy and cluster secondary ion mass spectrometry (cluster SIMS). Cluster SIMS provides the technology to determine the three-dimensional aspects of these devices and allows for a better understanding of drug delivery system structure and property relationships. These approaches will have long-term impact for product development in the pharmaceutical and biomedical industries, including the drug eluting stent manufacturers focused on new treatments for coronary artery disease.

Contact
Christine M. Mahoney

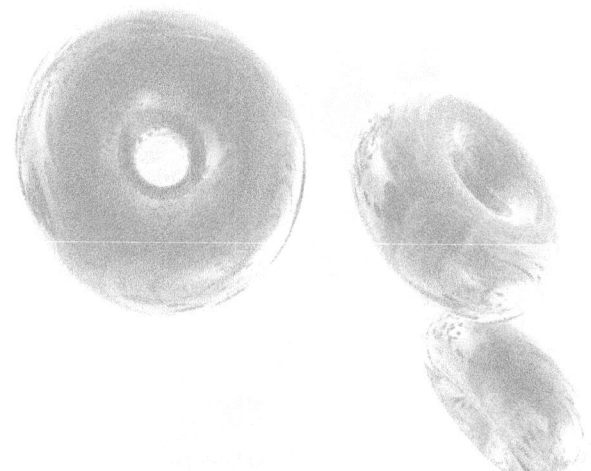

A POWERFUL TECHNOLOGY VIA A BEAM OF LIGHT

Finding the helper cells in the body of an HIV patient will enable a better fight against the deadly virus.

One of the most important laboratory tests performed on people infected with the Human Immunodeficiency Virus (HIV) is one that monitors the number of blood cells that the virus attacks and kills. It is these cells, the helper cells, that are responsible for helping to launch the normal immune response against microbial pathogens. Physicians use the results from the test to determine when to initiate therapies. If the helper cell number goes too low, then the patient can no longer fight off everyday infections and is likely to die. CSTL scientists are developing ways to calibrate the instruments used for helper cell measurements - the flow cytometer. One of the proteins in the helper cells protrudes through the membrane of the cell and is exposed on it surface. It can be detected by using a fluorescently tagged antibody that selectively only binds to the protein. The flow cytometer uses a laser to excite the fluorescent dye and counts the number of fluorescently labeled cells. CSTL scientists are working with international standards organizations to test potential reference materials and measurement procedures to directly address challenges that affect the accuracy and comparability of flow cytometer instruments. These efforts will enable better monitoring of the critical cells that can effectively fight off HIV in a patient.

Contact
Lili Wang

HEALTHY FOOD CHOICES LED BY ACCURATE LABELS

Measurement and standards related to nutrients in food products support compliance with relevant legislation.

Laboratories in the food testing and nutrition communities look to CSTL for methods and well-characterized certified reference materials that are food-based. These are necessary to facilitate industry compliance with labeling laws and improve the accuracy of information provided on product labels. CSTL researchers are working to provide food-matrix reference materials to facilitate compliance with nutritional labeling laws, provide traceability for food exports, improve the accuracy of label information for packaged foods, and contribute to studies of human nutritional status. Approximately 30 food-matrix materials are available to support these measurement needs; examples include infant formula, baby food, meat homogenate, fish tissue, baking chocolate, peanut butter, spinach, and food oils. Efforts are additionally focused on materials that are representative of foods with a broad range of fat, protein, and carbohydrate compositions. The SRMs support compliance with a number of federal regulations enforced by the U.S. Food and Drug Administration and the U.S. Department of Agriculture and contribute to reliable nutrition information on which consumers can base their dietary choices.

Contact
Lane C. Sander
Katherine E. Sharpless

QUALITY SUPPLEMENTS FOR HEALTH AND NUTRITION

Congress has recognized the lack of publicly available, validated analytical methods for dietary supplements.

Many consumers believe that botanical dietary supplements will improve their health and that these "natural" remedies are both effective and free from the side effects that may occur with other medications. There are occasional reports of inaccurate labeling, adulteration, contamination (with pesticides, heavy metals, or toxic botanicals), and drug interactions. In conjunction with the National Institutes of Health's Office of Dietary Supplements and the U.S. Food and Drug Administration, CSTL researchers are developing Standard Reference Materials (SRMs) for botanical dietary supplements. SRMs with assigned values for concentrations of active and/or marker compounds and toxic elements are being produced to assist in the verification of manufacturers' label claims and for use in quality control during the manufacturing process. Suites of materials based on green tea, black cohosh, blueberries, bilberries, cranberries, soy, kudzu, and red clover are currently in progress. A mixture of vegetable oils with values assigned for tocopherols is also being prepared as an SRM, as is a multivitamin/multielement tablet. This program provides the dietary supplement industry and measurement communities with the needed tools to improve the quality of dietary supplements and reduce public health risks that could be associated with these products.

Contact
Lane C. Sander
Katherine E. Sharpless
Stephen A. Wise

LINKING OCEAN HEALTH TO HUMAN HEALTH

Sentinel species such as whales, dolphins and other marine organisms provide insight to how the ocean environment affects people.

Due to their long life span, feeding at a high trophic level, and extensive fat stores that can act as a deposit for anthropogenic toxins dolphins have been proposed as a excellent organism in aquatic and coastal environments to monitor and understand the impact of infectious diseases, anthropogenic pollutants, and harmful algal blooms and their toxins. Through collaborative work with the Medical University of South Carolina, CSTL researchers are investigating the unique physiological adaptation of the dolphin lungs that could have therapeutic application to human related respiratory disease by making use of quantitative imaging techniques through the examination of how dolphin lung cells respond to low oxygen conditions and toxins. By developing methods to comprehensively characterize the growth characteristics of dolphin lung cells and compare morphological and image-derived data with dolphin lung cell gene expression profiles, the metrology infrastructure required for cell and tissue-based screening of marine mammal health is on the rise. This framework will enable linkages between ocean health and human health to be established.

Contact
John T. Elliott
Anne L. Plant

PAST INDICATORS FOR FUTURE PREDICTIONS

The long-term preservation of environmental specimens for deferred analysis and evaluation lends an eye to view and predict health trends.

Formal environmental specimen banking is recognized internationally as an integral part of long-term environmental health monitoring. The Marine Environmental Specimen Bank, established by NIST in 2001 at the Hollings Marine Laboratory in Charleston, South Carolina, cryogenically banks well-documented environmental specimens collected as part of marine research and monitoring programs directed toward understanding and predicting how the condition of our Nation's waters and coastal environments affect human health. Specimens include marine mammal tissues, mussels and oysters, fish tissues, seabird eggs, and peregrine falcon eggs and feathers. Many of these specimens are being analyzed retrospectively to determine time trends in emerging contaminants of concern in the environment and their potential time-line of exposure to humans. This bank is an important resource of research materials for documenting geographic and temporal trends in "new" pollutants, changes in transport and accumulation of "old" pollutants in the environment that might even be related to climate change, and for studying temporal changes in marine animal health through application of future new analytical and biochemical techniques.

Contact
Paul R. Becker
Rebecca S. Pugh
Stephen A. Wise

MATERIALS TO BETTER ASSESS THE HEALTH OF THE U.S. POPULATION

CSTL provides SRMs that are representative of the analytical challenges that may be encountered by analysts in all segments of the biomonitoring community.

For the past 25 years NIST has developed a number of biological fluid and tissue Standard Reference Materials (SRMs) primarily for clinically important analytes (e.g., cholesterol, glucose, creatinine, and trace elements) and contaminants. These materials support large scale biomonitoring programs such as those conducted by The Centers for Disease Control and Prevention and the New York State Department of Health along with other state agencies that require the analysis of a large number of samples obtained from study subjects. CSTL researchers and CDC are collaborating on the development of a wide variety of SRMs. These include organic contaminants in human serum, human milk, and human urine, lead in caprine blood, elements in bovine liver and animal serum, arsenic species in human urine, and toxic elements in human urine. Methods have been developed at NIST and at CDC for these measurements, and the results from the methods are being combined to provide certified concentration values for contaminants in the different materials. The development of these materials and methods greatly supports measurements for contaminants in biological fluids and tissues and expands quality assurance capabilities to important species critical to assessing the state of human health in the U.S.

Contact
Stephen A. Wise
Gregory C. Turk
Michele M. Schantz

Fuel & Feed
Advanced Energy Research Activities

Although America's appetite for energy appears boundless, traditional supplies are not. Promising alternative energy sources include biofuels, hydrogen fuels for automobiles, and photovoltaics. CSTL research activities in the energy sector seek the technology and innovation necessary to make alternative energy systems viable. Standards for next-generation fuels, based on previous approaches for traditional processes and feedstocks are needed. Pushing the bounds to permit biofuels of consistent quality, metrology and technology for large-scale hydrogen use, molecular structures that form the key components in solar-to-fuel technologies, and information underpinning advanced energy systems are just a few examples of CSTL's drive to address the nation's critical energy needs.

THERMOPHYSICAL PROPERTIES NEEDED TO GROW BIOFUELS INDUSTRY

Liquid fuels derived from biological materials need a close look as they become a part of our next generation energy solutions.

Transportation fuels will undergo major changes in the coming years, and liquid fuels derived from biological materials -- biofuels-- have very bright prospects. Understanding the thermophysical properties of biofuels is necessary if these are to be efficiently designed, produced, distributed, and enabled for use. These properties include the thermodynamic properties, such as density, heat capacity, and enthalpy, and the transport properties, such as viscosity and thermal conductivity. There is presently no comprehensive, consensus source of reliable property data for use by industry, thus limiting widespread and efficient adoption of biofuels. CSTL researchers are closing this gap through the development of thermophysical properties databases for biofuels. A combination of reviewing, modeling, and measuring activities is underway in parallel for the five most important fatty acid methyl esters that comprise biodiesel derived from vegetable oils. The findings offer better knowledge of fuel properties which will lead to better designs and shorter development times of everything from production facilities to emissions control systems for biofuels, fuels for which demand is expected to grow globally 20 percent annually through 2011.

Contact
Thomas J. Bruno
Marcia L. Huber
Mark O. McLinden

BETTER BIOFUEL MEASUREMENTS TO ENSURE QUALITY FUEL AND FAIR TRADE

Biofuel reference materials that are characterized for many of the properties of importance to various global producers will be critical for assessing biofuel quality.

With concerns over the limited supply of non-renewable forms of energy, renewable sources of energy such as biofuels are being investigated. The change to renewable energy, however, presents new questions as engines are modified, and many properties of biofuels change with environmental conditions. In conjunction with organizations within the European Union (EU), Brazil, and U.S. stakeholders, CSTL researchers are developing analytical methods for the properties of interest in biofuels and reference materials to support biofuel measurements. Two biodiesel reference materials, one prepared from soy and one prepared from animal fat, and two bioethanol samples, one prepared from anhydrous ethanol and one prepared from hydrated ethanol, are currently being analyzed by NIST and the National Metrology Institute of Brazil for a wide range of parameters such as fatty acid composition, sulfur and water content, and viscosity. An additional collaboration will involve the National Metrology Institute of the EU to develop biofuel reference materials from products commonly used within Europe and to assist in a laboratory proficiency testing scheme to ensure quality and comparability in biofuel measurements across the US and Europe.

Contact
Stephen A. Wise
Michael M. Schantz

Testing the Fuel of the Future

Fair hydrogen retail trade will require strict flow measurements to ensure the proper quantity of fuel is dispensed.

Hydrogen-fueled demonstration vehicles are refueled from dispensers in 3 to 5 minutes using sequential releases from a bank of pressurized cylinders. This results in large gas flow (0 to 10 kg/min), pressure (0 to 70 MPa), and temperature transients at the device metering the total hydrogen dispensed. To ensure equity in retail trade, regulators seek < 2 % uncertainty in the total hydrogen quantity delivered. However, hydrogen dispenser manufacturers have found errors of 8 % to 20 % in point-of-sale flow meters. CSTL investigators are examining the refueling process, using critical flow nozzles as working standard flow meters in a hydrogen flow system that mimics the transient conditions occurring during vehicle refueling. A thermodynamic model for the cylinder discharge and system design has been completed and evaluation of sensor time response has begun. This effort forms the basis for evaluating the performance of dispensing flow meter technologies and that of prototype field standards used by state inspectors to periodically test retail hydrogen dispensers. The needs of the vehicle fuel of the future will be met, a fuel that will reduce the dependence of the nation's personal transportation system on imported oil and minimize harmful vehicle emissions.

Contact
John D. Wright
John J. Hurly

HOMING IN ON ABSOLUTE HUMIDITY FOR THE ZERO EMITTERS

Calculated values of absolute humidity are needed for accurately measuring the water vapor content of hydrogen fuel.

CSTL researchers pursuing reliable humidity measurements and control of water content needed for the efficient operation of proton exchange membrane fuel cells, a type of fuel cell being developed for transport applications as well as for stationary and portable fuel cell applications with low to zero emissions. Thermophysical properties of water-hydrogen mixtures over elevated operating pressure (up to 100 MPa) and temperature ranges have been calculated. Second virial coefficients, needed to predict departures from ideal behavior, have also been calculated for water-air and water-hydrogen mixtures. Using these results absolute humidity, has been determined for this pressure range and for saturation temperatures over the range -70 °C to 90 °C. These findings will form a basis for the NIST humidity standard of compressed hydrogen, oxygen and air for calibration of hygrometers used in measuring water vapor content of hydrogen fuel in the fuel cell industry, an industry that speaks to a global market projected to reach $2.6 Billion in 2009.

Contact
Peter H. Huang

ON THE ROAD TO COMMERCIALLY VIABLE HYDROGEN-POWERED FUEL CELLS

A detailed understanding of the elemental chemistry and impurities that may enhance or diminish hydrogen storage capacities is necessary to accelerate the development of hydrogen fuel cells.

No existing hydrogen storage materials satisfy the U.S. Department of Energy's goal for achieving a commercial hydrogen storage system for competitive hydrogen-powered transportation. Today's methods of material synthesis generate only test quantities that require rapid chemical and impurity analysis, preferably by nondestructive techniques. CSTL researchers are investigating analytical methods suitable for the task. Prompt gamma activation analysis (PGAA) and instrumental neutron activation analysis (INAA) are being used to determine the essential elements in hydrogen fuel chemistry such as hydrogen, boron, and magnesium. Determining these elements by any other nondestructive techniques is neither trivial nor reliable. In addition, CSTL researchers are developing standards and analytical methodology for compounds that must be measured to ensure fair trade such as hydrogen sulfide and ammonia. Compounds such as these have been identified as detrimental to the performance of the hydrogen cell. These efforts are part of the larger federal endeavor to make hydrogen fuel an economical and safe alternative to carbon-based fuels.

Contact
R. Gregory Downing
Leonid A. Bendersky

SCIENTISTS LINK STRUCTURE AND DYNAMICS TO SOLAR DEVICE EFFICIENCY

Understanding electronic structure and ultrafast dynamics at interfaces in photovoltaics is necessary to bring these devices to market.

Many scientists and engineers believe organic solar cells will provide a cheaper alternative to traditional inorganic cells since large-scale production of organic polymers could be less expensive than the fabrication of inorganic materials such as silicon. However, organic solar cells have much lower efficiencies than traditional technologies. CSTL scientists are trying to raise the efficiency of organic photovoltaic device structures through the optimization of interfacial electronic structure and nanoscale morphology in model devices. Studies on correlations between key electronic and structural characteristics of organic donor-acceptor interfaces to elucidate the factors that control the efficacy of charge separation, a critical factor to raising organic device energy efficiency, are underway. Development of photoelectron spectroscopies, scanning probe methods and theoretical models of multiscale processes in molecular materials for thin organic films and heterojunctions of materials are core to this pursuit. These findings could substantially increase understanding the processes that will make solar electricity from photovoltaics cost-competitive with conventional forms of electricity in the utility grid, a challenge to be met by 2015 under the Solar America Initiative.

Contact
Steven W. Robey

LAUNCHING REUSABLE ROCKET TECHNOLOGIES

A robust properties model that is based on reliable e[xperi]**mental measurements is the best means to provide i**[nforma]**tion to rocket engine designers.**

Environmental concerns, the potential for disruptions in supply, an[d need] for reusable rocket engines have generated interest in new aviati[on and] rocket fuels. Accurate knowledge of thermophysical properties is a pr[erequisite] to the design of efficient and cost-effective engine systems that utiliz[e these] fuels. CSTL researchers are taking a combined experimental and mod[eling ap]proach to elucidate the behavior of key properties over wide ranges o[f tempera]ture and pressure. Measurements and models for several fuels, includi[ng the] rocket propellants RP-1 and RP-2, a natural-gas derived aviation fuel S[8, a coal]-derived liquid fuel, Jet-A, and JP-8. This research comprises measure[ments of] the density, speed of sound, viscosity, and thermal conductivity and r[epresent] the most accurate and extensive data available. The pure fluid models [and the] surrogate mixture models developed in conjunction with these measu[rements] will be incorporated into the next release of a standard reference dat[abase di]seminated to engineering and research communities worldwide, REFP[ROP. This] database provides an easy-to-use thermophysical properties package t[o enable] efficient and cost-effective engine system design and production.

Contact
Thomas J. Bruno
Marcia L. Huber
Mark O. McLinden

THE PATH TO "ONE FUEL FORWARD"

As our defense industry moves towards its single-fuel vision, new fuels and engines are being developed: matching fuel with engines requires chemical and physical property data.

A knowledge base of chemical and physical property information, and specifically how the two types of information are interrelated, is needed to meet major energy concerns with respect to resources, environmental responsibility, and national security. CSTL researchers, in collaboration with the Department of Defense, have developed a new method that is substantially contributing to this needed knowledge base, called the advanced distillation curve method, and have applied the technique to rocket propellants, gasoline, jet fuels, diesel fuels (including oxygenated diesel fuel and biodiesel fuels) and crude oils (including a crude oil made from pig manure). The method provides the relationship between the thermodynamic description of complex-fluid volatility and the composition of the fluid. These data are necessary for the design and testing of new fuels. The U.S. Air Force (USAF) has used these measurements in its One Fuel Forward initiative, in which a single fuel is desired for aircraft, trucks and fixed stationary power equipment. In addition, in its intent to find alternative fuels for aircraft, USAF demonstrated the flight of a B-52 aircraft on a mixture of conventional JP-8 and a synthetic called S-8. The choice and operating specifications were made in part on the basis of the CSTL advanced distillation curve data. Current and future fuel design and specification will rely heavily on this effort.

Contact
Thomas J. Bruno
Tara M. Lovestead

REALISTICALLY SIMULATING COMBUSTION FOR BETTER TRANSPORTATION

Future engines will require knowing the complex interactions of the combustion chamber with the chemical and physical properties of fuels.

Efficiency of transportation engines directly affects the economy, and emissions of CO_2 and other pollutants have significant implications with respect to climate change, health care, and the environment. Accurate combustion-models are needed to enable advanced computer based strategies for the design of more efficient and less polluting engines, and engines that take advantage of alternate fuels. At present the physical and chemical data necessary to develop these combustion models are not available. With support from the U.S. Air Force and the Department of Energy, research aimed at determining this type of information is underway in CSTL. Scientists are obtaining new data through unique experiments in a shock tube laboratory. Experimental kinetic data are being used to fill data gaps and resolve uncertainties in key data essential to combustion models, and to derive fundamental structural activity relationships to predict the behavior of components in real fuel mixtures when combusted. A database has been established to provide a centralized site for combustion model information and employs heretofore lacking standards for nomenclature, notation, and traceability. This work is facilitating coordinated, collaborative research for the design of innovative power systems and assures U.S. engine manufacturers a basis for innovation and international competitiveness.

Contact
Wing Tsang
Jeffrey A. Manion
Donald R. Burgess
W. Sean McGivern

REDUCING CO$_2$: NEW POWER PLANTS NEED DATA ON MOISTURE IN GASES

The amount of water in the vapor phase in some novel electric power plant technologies is important for the process economics, but experimental data are lacking and various modeling approaches attempted by industry have not given consistent results.

Thermophysical properties of gas mixtures containing water are important in several contexts, such as humidity standards and the design of advanced power cycles for electricity generation. In collaboration with researchers at the University of Nottingham in the United Kingdom, and with support from the U.S. Department of Energy and the Electric Power Research Institute, CSTL researchers are using computational approaches to obtain second virial coefficients (a measure of gas-molecule-to-gas-molecule interactions that deviate from the ideal-gas behavior) between water and common gases over wide temperature ranges (and with lower uncertainty than most existing experimental data). The new results are being applied to air-water mixtures for humidity standards, to the equilibrium of synthesis gas (CO and H$_2$) with water, and to the thermodynamic properties of a suite of combustion gases. Efforts are underway to confirm model calculations with high-temperature experiments, and to measure the thermal conductivity of moist gas mixtures. This program will help enable a promising technology being developed for generating electricity from coal with minimal CO$_2$ emissions and speaks to the growing needs of designing turbines and other equipment that make use of gaseous combustion processes.

Contact
Allan H. Harvey
Mark O. McLinden

PLUGGING-IN TO CLEANER ENERGY

To demonstrate compliance with environmental regulations, industry will require methods and standards along the full product lifecycles of coal, oil, and natural gas, which account for over 85 % of the energy consumed by the U.S.

Along with the need for more efficient utilization of fossil fuels, there is a need for the environmental management of combustion emissions and waste products, and pollutant emissions from vehicles using liquid fuels. Sulfur, mercury, and carbon are subject to environmental regulations. Management of energy production from the processing of raw fossil fuel to the release of environmental pollutants from combustion processes, together with risk assessment of waste products requires accurate analytical measurements and relevant reference materials. Established 40 years ago, our fossil fuel SRM program with a current inventory that includes coals, cokes, residual fuel oils, distillates, and gasoline. CSTL researchers are increasing the availability of these matrix-specific reference materials to support current and future measurement needs of petroleum industry and the fossil fuel-based electric utility industries, and are developing new analytical methods to support new or enhanced standards for priority areas such as high-accuracy carbon fossil fuel standards for assessment of CO_2 emission inventories and trading. In addition, CSTL is providing primary standards for the electric utility industry to comply with regulatory requirements for mercury emissions monitoring and reduction. These efforts will enable the current fuel sectors to meet life-cycle assessment challenges in a cost effective manner with the least disruption in product availability.

Contact
Stephen E. Long
William R. Kelly

Safe & Secure
Public Safety/Security Research Activities

CSTL plays a key role in enhancing the nation's homeland security. Through the development of measurements and standards infrastructure that ensures the accuracy, reliability, and security of systems critical to public and national security, CSTL provides products and services that help Federal, state, and local organizations meet their responsibilities to protect the lives, rights, property, and privileges of the members of the U.S. public. CSTL researchers develop, compare and test new technologies to identify national threats and put forth the standards necessary to ensure safe and responsible responses to incidents. Many of these technologies and products are now used internationally, ensuring mitigation of threats prior to their arrival at U.S. borders.

GUIDELINES TO KILL BIOLOGICAL THREATS

Biofilms offer a potential reservoir for dangerous organisms and also serve to protect these organisms from disinfectants, although their role in such an event has not been quantified.

Biofilms are communities of microorganisms attached to surfaces that predominate in water/surface interfaces common to nearly all ecosystems. CSTL researchers, with support from the U.S. EPA National Homeland Security Research Center, have developed an experimental pipe system to grow naturally occurring biofilms using well-characterized synthetic tap water in order to investigate the disinfection process necessary to kill a harmful biological outbreak. After establishing healthy biofilms, the system was subjected to simulates for real biological threats such as, **_Bacillus anthracis_**, ricin, **_Escherichia coli_**, and the bacteria that is the cause of tularemia. Initial studies focused on the use of bleach to inactivate the spores attached to the pipe surfaces and in the water. These were extended to additional disinfectants and the fate of these threats and the efficiency of decontamination in large-scale water systems were studied. Using a unique method to sensitize the spores to disinfectants, the scientists developed a method to remove spores from biofilms without the use of large quantities of dangerous chemicals. This work is part of a larger project to study the safety of building water systems in collaboration with researchers in NIST's Building and Fire Research Laboratory. Guidelines are being developed that will aid in the decontamination of water systems in buildings in the event of contamination and to ensure the safety of our nation's drinking water.

Contact
Kenneth D. Cole

DETECTING HIGH EXPLOSIVES FROM THEIR GAS

Accurate detection of explosive compounds in airports and at ports of entry around the globe requires device certification.

Explosive compounds need to be detected on a variety of surfaces - clothing, suitcases, shoes, just to name a few. Every surface will interact with the compounds and the degree of adhesion will vary, surface to surface. By and large, detection of these compounds relies on getting the molecules off the surface and into the gas phase. While there have been significant advances in instrumentation for in-the-field applications, a device to be used for airport security will require certification by federal authorities to ensure accuracy. However, the lack of fundamental data to understand the vapors of explosives, those which are detected by field devices, is hindering approaches to certification. With support from the Department of Homeland Security, CSTL researchers are conducting studies to fill this gap by elucidating the vapor behavior of solid explosives. These studies are providing measurements of the vapor pressures related to energetic materials, including the taggents 2-nitrotoluene, 3-nitrotoluene, and 4-nitrotoluene, used in high explosives. A knowledge base of the temperature dependent composition of the vapor above high explosive materials and of the surface energetics of the interaction of explosive vapors with surface substrates is being created. This work is providing the required measurement infrastructure to certify in-the-field detection devices and is leading to the development of next generation explosive measurement technology.

Contact
Thomas J. Bruno
Jason A. Widegren
Tara M. Lovestead

SMALL PARTICLES ENSURE SAFE SENSING OF DEADLY THREATS

Standard test particulate materials are needed to verify the performance of trace explosives detection portals, technology used for rapid screening of people for high explosives.

The preparation of realistic standard test materials for explosive portal systems, such as those in airports, requires fabrication of particles of known size and compositions deposited onto an appropriate substrate in well characterized numbers. One intriguing method for generation of such test standards is the production of uniform polymer microspheres containing explosives or explosive stimulant compounds of interest. Polymer microspheres are attractive for this application because their size and composition may be specifically tailored to testing portal system specifications. Another advantage is a potential reduction in vapor pressure of the analyte compounds encapsulated in the polymer matrix which extends the useful lifetime of the standards. Prototype polymer microspheres are being constructed by CSTL researchers through the use of a novel sonicated annular co-flow device that delivers precisely controlled liquid droplets into water where they undergo an oil/water emulsion process and eventually cure into hardened microspheres. The research team has successfully incorporated several high explosives into a bio-friendly polymer matrix to provide a test material that is both safe with regard to human exposure and non-contaminating with regard to actual explosives being detected. These efforts are delivering the technology needed to produce critical materials for testing the reliability of thousands of trace explosive detectors deployed at airports by the Transportation Security Administration.

Contact
Matthew E. Staymates

MICROSCOPIC PAINTBALLS KEEP CONTRABAND DETECTORS IN-CHECK

Inkjet-based dispensing produces well characterized materials for verifying performance of trace contraband detectors.

Innovative chemical sensors are continually emerging to detect concealed illegal materials such as explosives, drugs, and chemical weapons on people, in luggage, and in shipping containers. Reference materials which closely simulate samples, such as swipes, are needed to critically test the performance of trace contraband detectors and support the continual improvement of detection technologies as threat priorities evolve. CSTL researchers are developing methods for depositing known quantities of contraband (and associated materials) onto the surfaces of swipes. These well-characterized swipes can effectively test the performance of trace contraband detectors. Using a commercial inkjet device that was adapted to enable precise spatial and quantitative deposition of trace materials in the form of picoliter droplets, a trillionth of a liter, onto a variety of surfaces, swipes of known composition are being generated. The composition of the swipes is verified using novel gravimetric and optical methods and quantitative deposition of picogram amounts, one trillionth of a gram, onto the surfaces of test swipes has been demonstrated. The well-characterized swipes will make it possible to verify contraband detector performance in the field, a critical step to enabling the acceptance of emerging technologies by the Transportation Security Administration.

Contact
R. Michael Verkouteren

PUTTING A STOP TO ILLICIT DRUGS AT THE BORDER

The reliability of field-deployable techniques to discriminate legal pharmaceuticals from illicit drugs is critical to stopping illicit drugs from entering our country.

Law enforcement is continually faced with discriminating harmless pharmaceutical products from illicit drugs. This is particularly acute at our borders, where a wide variety of tablets and powders are encountered. The fastest growing area of drug abuse in the U.S. is represented by illegally diverted pharmaceuticals, specifically hydrocodone (Vicodin), oxycodone (Oxycontin), and alprazolam. Illicit drug samples are also complex, containing multiple drugs and excipients. The reliability of Ion Mobility Spectrometry (IMS) to correctly identify illicit materials given such sample complexity was investigated. A comprehensive sample suite representing over 80% of drug samples reported by state and federal forensic laboratories over the past 7 years was developed for testing purposes. The selectivity of IMS for the drugs in this sample suite was evaluated, and specific guidelines were developed for optimizing IMS alarm responses. In addition, field-deployable sampling procedures for tablets and powders were developed. Test materials containing cocaine - produced by picoliter inkjet printing - have been analyzed by both CSTL and the Drug Enforcement Administration to assess the quality and comparability of IMS results. This program provides critical evaluations, data, and materials needed by local law enforcement, U.S. Customs and Border Protection, the U.S. Coast Guard, the National Guard, and federal and state correctional facilities to further their efforts to accurately detect and identify illegal narcotics.

Contact
Jennifer R. Verkouteren

NUCLEAR FORENSICS BOLSTERS NUCLEAR NON-PROLIFERATION

Rapid isotopic screening of a collection of particles provides a means to quickly determine whether a uranium enrichment facility is producing products outside of its declared limits.

Enriched uranium describes uranium in which the percent composition of uranium-235 has been increased through the process of isotope separation at specialized facilities. Enriched uranium is a critical component for both nuclear power generation and nuclear weapons. The International Atomic Energy Agency, under the auspices of international treaty, monitors enriched uranium supplies and processing to ensure nuclear power generation is decoupled from nuclear weapons production. CSTL supports the IAEA mission in its research to measure uranium by secondary ion mass spectrometry (SIMS). For uranium enrichment monitoring, researchers have demonstrated the rapid measurement of uranium isotope ratios in large numbers of micrometer-sized particles and pinpointed those that are inconsistent with the natural environment. The SIMS instrument offers high sensitivity with high mass resolution, and semi-automatically can examine a large number of uranium-bearing particles that are dispersed on a flat substrate. The primary ion current is adjusted to yield a measurement precision of about 1 % on the isotopic ratio for a typical particle. The isotopic measurements can be conducted as fast as 2.5 minutes per particle so 100 particles can be screened in about 4 hours. The screening of materials for perturbed isotopic ratios has potential in a wide range of scientific and forensics applications.

Contact
David S. Simons

DUAL-USE TECHNOLOGY FOR PUBLIC SAFETY AND HEALTH

Luminescence measurements have become the methods of choice for chemical detection and new clinical analyses due to their high selectivity and sensitivity.

First-responders to an accident or terrorist attack require equipment that can accurately determine what deadly chemicals agents might be present. CSTL scientists are helping manufacturers of this type of equipment with tools they can use to calibrate their products and ensure their accuracy and reproducibility. The most commonly used field instruments make use of technologies based on fluorescence. This involves the use of a laser that is shone on a fluorescent molecule to generate a signal of light of another color, the intensity of which corresponds to the concentration of the chemical detected. Fluorescence possess characteristics that make it superb for measuring very low concentrations of threat agents, but also require very precise calibration of the instruments. CSTL researchers have developed SRMs and reference procedures to enable calibration and validation of the way in which the measurements are performed. These include an atomic light source that mimics the signal generated by the fluorescent molecule and chemically-doped glass that emits precise wavelengths of color. The work is critical to national defense efforts and potentially has broad benefit for clinical diagnostics.

Contact
Paul C. DeRose
Steven J. Choquette

VERY SMALL COOLERS FOR SENSITIVE DETECTORS

As sensors evolve, becoming smaller with more compact arrangements, there exists a need to provide chip-scale cooling of the devices.

Cryogenic temperatures are required for many technology areas, including infrared and terahertz sensors for surveillance. Such systems could be used for detection of enemy missile launches (infrared), chemical identification (infrared and terahertz spectroscopy), or smuggling of nonmetallic weapons underneath clothing through airport security lines (terahertz imaging). The cool environment increases sensitivity by providing a low-noise environment for detectors. However, current state-of-the-art cryocoolers require at least several watts of input power and occupy tens of cubic centimeters of volume, limiting their applicability outside of the laboratory. They are designed for refrigeration powers of at least 150 mW. As detectors of various wavelengths of radiation improve, more useful detection can be accomplished with only 3 to 10 mW of refrigeration power at temperatures in the range of 77 to 150 K. Input powers would be less than 1 W, which can be provided by small batteries. These "microcryocoolers" can lead to useful hand-held detection systems. CSTL researchers, in collaboration with scientists from EEEL and the University of Colorado, are developing a mixed refrigerant Joule-Thomson cryocooler for terahertz detector cooling. Joule-Thomson cryocoolers are attractive for miniaturization, because the cooling is achieved through a simple throttling of the gas used in the refrigeration cycle, typically accomplished with small diameter tubing. Such developments will enable terahertz technologies for imaging applications critical to public safety and health.

Contact
Ray Radebaugh
Daniel G. Friend

MISSION-CRITICAL TOOLS FOR PUBLIC HEALTH AND SAFETY

CSTL develops critical chemical measurement methods and reference materials to support other federal agencies in their missions to ensure public health and safety.

CSTL has a long history of working with other Federal agencies to provide research, measurements, methods, and standards to support public safety and homeland security. Recent activities have been supported by the U.S. Departments of Justice, Homeland Security (DHS), Defense (DoD), and Energy (DoE); National Institute for Occupational Safety and Health (NIOSH); and Consumer Product Safety Commission (CPSC). Many of these activities are coordinated through the NIST Office of Law Enforcement Standards. For over 20 years CSTL has provided SRMs to support forensic analyses related to drugs of abuse including alcohol. These SRMs provide critical calibration of instrumentation where quantitative analysis directly affects criminal prosecutions. CSTL also provides numerous SRMs and RMs for drugs of abuse in urine and hair. A serum SRM will be issued in 2009 with values assigned for seven drugs, including two (methadone and nordiazepam) that have not been measured in previous matrices. In recent years CSTL has been providing support to DHS in the area of standards and measurements for trace explosives detection. Material matrices include a smokeless powder to assure quality of 'low explosives' type measurements. The first SRM for high explosives will be available in 2009. A beryllium oxide SRM has also been developed in collaboration with DoE and NIOSH to support measurements of potential aerospace and defense industry worker exposure to beryllium. A new project with the CPSC this year will develop tools for the validation of test methods for lead in paint on children's products.

Contact
Stephen A. Wise
Lane C. Sander
Gregory C. Turk

ELECTRONS & VOICE
Electronics and Telecom Research Activities

Electronics and communications are essential to our everyday lives. Cutting edge CSTL research is providing dramatically improved measurement tools, models, and data to keep U.S. electronics and telecommunications R&D innovative and globally competitive. These outputs are vital to enabling next generation electronic devices, such as organic thin film transistors, future semiconductor device fabrication, and superior telecommunications for U.S. citizens and our military. Research activities consider devices, circuits and electronic systems. Microelectronic investigations are a particular emphasis, as these provide the industry with effective tools for future improvements in micro and nano-technologies, intelligent sensors, electronics for telecommunications, digital system electronics and electronics for process control.

THE PERFORMANCE OF NEXT GENERATION ELECTRONIC DEVICES

Organic thin film transistors have the potential to become the replacement technology for the semiconductor-based electronics we use today.

Thin film electronic devices fabricated using organic semiconductors, insulators and conductors are expected to greatly impact the semiconductor electronics industry by significantly reducing manufacturing costs and expanding the applicability of active thin film electronics. Despite significant improvements to the electrical performance of organic thin film transistors (OTFT) over time, surprisingly little is known about the fundamental mechanisms governing charge injection and transport in the transistor channel. In collaboration with scientists from the University of Kentucky and the Pennsylvania State University, CSTL researchers are using Scanning Kelvin probe microscopy to correlate film microstructure with device performance for thin film transistors. This approach provides a platform to simultaneously probe the topography and the local potential drops within working devices. A correlation between organic thin film microstructure and potential drops within the working device has been demonstrated. This work illustrates the importance of processing and measurements in the development of optimal low cost, organic electronic devices, a market predicted to achieve $4.9 billion in revenues by 2015 as needed improvements in electron mobilities, switching speeds and environmental stability are attained.

Contact
James G. Kushmerick

TOOL TO STUDY ALTERNATIVE METALS FOR BETTER CHIPS

Nanoscale calorimeters can provide critical quantitative data needed to support the development of advanced materials for better semiconductor device performance.

Growth in the semiconductor industry is driven by the need to enhance performance by packing more transistors into the same chip area. As the feature size shrinks, the limits in performance of conventional materials are being reached, thereby limiting growth for this market. The industry is moving towards the use of fully silicided metal gates for the metal layer. An important example is the nickel-silicon intermetallic system, which is made by thermally annealing separately deposited nickel and silicon. The intermetallic phase or phases that form depend on the ratio of the two elements and on the annealing process. Further, the thermal annealing process must be compatible with other processes that are used in making the semiconductor devices on a wafer. This favors rapid and ultrarapid processes. CSTL researchers are using novel nanoscale calorimetry to characterize rapid thermal processes with heating rates inaccessible to conventional calorimeters. It has been shown that at very rapid annealing rates, one phase is bypassed during the ramp, and that a metastable liquid is formed that enhances the mixing of nickel and silicon. This measurement technique will provide the semiconductor industry with a tool to investigate new materials essential for better and faster devices and ultimately market growth. The current global market for semiconductor chips is on the order of $270 billion.

***Contact
Richard E. Cavicchi***

NOVEL METHOD TO MEASURE THE PHYSICAL PROPERTIES OF VOLATILE LIQUID PRECURSORS

Key characteristics of organic films in semiconductors show crucial dependence on growth conditions such as the precursor partial pressure and flow rate.

In 1965 Intel co-founder Gordon E. Moore noted that since the invention of the integrated circuit in 1958, the number of transistors that can be placed inexpensively on an integrated circuit increased exponentially, doubling approximately every two years. This trend, Moore's Law, continues through today and is expected to do so well into the future. In striving to adhere to Moore's Law, the semiconductor industry is designing and using new metal-organic compounds, metals in combination with carbon-based compounds such as conductive polymers, plastics, or small molecules. Volatile liquid precursors compounds deliver a metal atom to the surface of a hot silicon wafer in many film deposition processes; the physical property most important for efficient delivery is vapor pressure because it controls the behavior of vapor delivery devices. A team of CSTL researchers is producing accurate vapor pressure data for relevant liquid precursors from room temperature to as high as 200 °C using pressure gauges that operate at the same temperature as the sample. This direct method was chosen over indirect methods that rely on, for example, thermogravimetric analysis or nuclear magnetic resonance. To measure thermal stability of precursors, a gas chromatograph mass spectrometer is incorporated into the apparatus. Such data will improve the modeling of chemical vapor deposition processes and the design and use of bubblers and other devices that deliver precursors to wafer surfaces for the semiconductor industry.

Contact
Robert F. Berg

MEASUREMENTS AND MODELS TO ADVANCE PLASMA PROCESSING

The semiconductor industry requires improvements in plasma etching and deposition process to fabricate future generations of devices.

Plasma processes and equipment face increasingly stringent requirements due to the need to maintain high device yields at decreasing feature sizes, the introduction of new materials, and the constant pressure to keep production efficiency high. To meet these challenges, industry requires better, more predictive modeling of the impact of plasma equipment on process results. Industry also requires the development of robust sensors and process controllers which are able to convert large quantities of raw data into information useful for improving manufacturability and yield. In collaboration with researchers from the Korea Research Institute of Standards and Science, CSTL researchers are developing a variety of measurement techniques that provide industry and academia with methods to characterize the chemical, physical, and electrical properties of plasmas. The techniques include improved laboratory diagnostic measurements and robust, non-perturbing measurements for use in process monitoring and control in manufacturing. Experimental data are being measured under well-defined conditions in highly-characterized standard plasma reactors. These data are applied to the development of models needed to gain insight into complex plasma properties and behaviors. We rigorously test and validate the models by comparisons with experimental data. Together, the measurement techniques, data, and models are assisting the semiconductor manufacturers and plasma equipment manufacturers to better control existing processes and tools and help them to develop new ones.

Contact
Mark A. Sobolewski

CRYOCOOLERS FOR DIGITAL COMMUNICATIONS

An analytical model has been developed to design cryocooler systems for sensitive communication technology.

All-digital radio communication between the various forces of the U.S. military is required in the near future. Such communication requires analog-to-digital converters of speeds only possible with low temperature superconducting electronics. The replacement of existing rack-mounted rf electronic communication equipment with superconducting digital equipment requires the development of compact 4 K cryocoolers that can provide about 0.1 W of refrigeration at 4 K with much less than 1 kW of input power. Existing 4 K cryocoolers require at least 1 kW of input power, and are too large and heavy for military operation. CSTL researchers have determined that major sources of entropy generation in such cryocoolers are those in the 4 K regenerative heat exchanger (regenerator) and the 4 K pulse tube component. The real-gas effect in the ^{4}He working fluid is also a major source of loss in these 4 K cryocoolers. As such, the researchers seek to determine how to increase the efficiency of the 4 K stage by a factor of about four while still being able to operate at a relatively high frequency of 30 Hz, so that the compressor can be made compact. Changing to a ^{3}He system may be a key improvement. In collaboration with researchers from Zhejiang University in China, an equation of state and transport equations for ^{3}He have been incorporated into the new NIST software known as REGEN3.3. The software was made publicly available in March 2008 on the NIST web site and will be essential for meeting the communication challenges of the future.

Contact
Ray Radebaugh
Daniel G. Friend

SMALLER & SMALLER
Nanotechnology Research Activities

U.S. industry is the largest consumer of CSTL's products and services. These customers come from established industrial sectors, such as the chemical manufacturers, and emerging industries, such as nanotechnology. Measurements at the nanoscale are critical to underpin progress toward nanotechnology's many anticipated market applications and societal benefits worldwide, from cost-competitive solar power to reliable supplies of clean drinking water. CSTL provides innovation to chemical, device, and process characterization at this scale, down to assemblies of a few atoms to materials in simple and complex media. CSTL nanotechnology research activities support all phases of nanotechnology development, from discovery to production.

SUPERRESOLVING MICROSCOPY FOR ANALYSIS AT THE NANOSCALE

The utility of superresolution microscopy is being extended by combining it with spectroscopy to achieve superior images of nanoscale materials.

Many of the technologies earmarked for significant growth in the coming decades are critically dependent on credible nanometer-scale measurements and analysis. These certainly include biomedical applications, high-density electronics, biosensors, high throughput experimentation, separations, and catalysis. Development of robust, nanoscale measurement tools is critically important if the full potential of these technologies is to be realized. Light microscopy is a widely used analytical tool because it provides non-destructive, real-time, three-dimensional imaging with chemical and material specific contrast. However, today's demands on chemical imaging have grown beyond current capabilities. Of particular interest is the extension of resolution enhancement to contrast mechanisms based on intrinsic chemical contrast, like the spectroscopic fingerprint that spontaneous and coherent Raman scattering can provide for non-labeled samples. A team of CSTL researchers has designed and fabricated a flexible superresolution optical microscopy which combines structured-illumination techniques with Raman-spectroscopy to record 100 nm resolution images with chemically-specific contrast. This platform represents the state-of-the-art in optical diagnostics for in-situ characterization of organic and biological materials and will accelerate progress in the development of nanotechnology.

Contact
Stephan J. Stranick

MEASUREMENTS OFFER A VIEW TOWARD DEVICE DEVELOPMENT

Characterization tools are being developed to understand how zinc oxide nanowires grow, a process difficult to elucidate by just one technique.

Zinc oxide (ZnO) is one of the most important nanomaterials for nano-optoelectronics, sensors, transistors, and nanopiezotronics. Because of the unique piezoelectric and semiconducting dual properties, ZnO nanowires (NWs) are the fundamental material for nanogenerators, which convert mechanical energy into electricity. However, current uses are limited by the ability to produce high quality nanowires. Copper is an attractive catalyst for generating zinc oxide nanowires, although defects due to impurities, oxygen deficiencies, and structural defects lead to decreased optical and electrical transport. CSTL researchers are investigating novel techniques to accurately characterize ZnO NWs grown on bulk copper by chemical vapor deposition. Photoluminescence microscopy, high-resolution transmission electron microscopy, energy dispersive x-ray spectroscopy mapping and auger electron spectroscopy are providing insight to ZnO NW growth and development. The latter two techniques have clearly shown copper throughout the length of the nanowire when grown, which is most likely responsible for the strong emission from these nanowires. Auger electron spectroscopy also has revealed large amounts of oxygen on the surface of these NWs, a further contributor to defect emission states. These protocols will enable the nanotechnology industry to better understand the key factors in ZnO device fabrication and performance and will enable ZnO applications to be realized.

Contact
Steven A. Buntin

LAB UNDER CONSTRUCTION: APPROACHES TO REALIZE REVOLUTIONARY LABS-ON-A-CHIP

Horizontal nanowire formation is being used as a template for developing nanochannels with specific size limits, an essential tool for lab-on-a-chip development.

Controlling the location and physical orientation of nanobuilding blocks in assemblies of nanoscale devices become challenging tasks when it comes to the use of "bottom-up" chemical approaches for nanodevice fabrication. CSTL researchers have developed a strategy for *in situ* fabrication of horizontal nanochannels, which allows control over their orientation and average pore size to an unprecedented range of 5–20 nm. Using this approach, selectively fabricated nanochannels with controlled average pore sizes at known positions on a sapphire surface have been created. Equally important, nanochannels, either single or group, are addressable photolithographically and integrable to microchannels without the use of any high resolution lithography. This technique uses horizontal ZnO nanowires (NWs) as a sacrificial template. Cross-sections of nanowires and nanochannels are prepared from this template via a focused-ion beam and examined by electron microscopy and energy-dispersive X-ray spectroscopy. This approach could provide the capability of fabrication of complex micro- and nanofluidic platforms, tiny labs-on-a-chip. Precise design of the channels on these chips is essential to advancing their application in nanomedicine, energy conversion and storage, agriculture and food systems, realistic multiphenomena/multiscale simulations, and environmental applications.

Contact
Babak Nikoobakht

ENABLING OPTICAL TECHNIQUES TO STUDY NANOSCALE FILMS

Methods are under development to characterize critical molecular transformations during thermal processing of organic semiconductor films.

A wide range of technologies are impacted by the performance of thin (<100 nm) layers of advanced materials. This is particularly true in the emerging macroelectronics industry and in biomaterials. Rational development of these materials is hampered by the lack of structure-function relationships due to a severe lack of adequate data/measurement tools for molecular level characterization of thin films. The traditional structural tools of x-ray diffraction and NMR are severely limited due to lack of signal and contrast. CSTL researchers are developing and demonstrating the ability of bench-top optical tools, combined with advanced physical models, to determine the details of molecular structure in thin films and at interfaces. Specific emphasis is placed on polarized photon spectroscopies: spectroscopic ellipsometry, infrared absorption, sum frequency generation, and coherent anti-Stokes Raman scattering. When possible, measurements are complemented and validated by comparison to unique NIST resources such as neutron scattering and synchrotron-based x-ray techniques. Specific recent technical advances have been demonstrated through the analysis of polymer semiconductors. Polymer semiconductors are inexpensive solution processable alternatives to amorphous silicon for applications in flexible large area electronics. The findings will enable rational design of future highly ordered polymers for electronic applications.

Contact
Lee J. Richter

TESTING ONE OF THE MOST POWERFUL ELECTRON MICROSCOPES

The need for characterizing the atomic-scale structure of nanoparticles is highly important for understanding their behavior and optimizing their synthesis.

Subtle changes in chemistry, especially at the very surface of nanoparticles can have major impacts on the properties which they subsequently exhibit. In order for the development of true structure-property correlations to be realized for these materials, careful measurement of the position and identity of the various constituent atoms becomes essential. CSTL researchers are developing approaches to use the revolutionary Titan™ 80-300 S/TEM (scanning/transmission electron microscope) to carry out high spatial-resolution characterization of the CdSe/ZnS quantum dot structure in a compositionally sensitive mode known as high-angle annular dark-field imaging. Since the current in the electron probe in this instrument is sufficient to significantly alter or even destroy the underlying material, it is necessary to limit the dose imparted to the sample in order to prevent structural alterations from occurring during the analysis. To this end, a time-series approach to imaging these quantum dots has been employed. This involves the collection of several, low-dose (i.e., fast scan) images from the same area of the sample and then correcting any drift between them. The individual images are inherently noisy due to the low signal they contain; however, the summation of the entire series produces an atomic-scale resolution image with superior informational content. This demonstrated atomic-scale, compositional characterization of nanoscale structures will lead to protocols to discover new relationships between structure and properties, relationships that are driving development at the nanoscale.

Contact
Keana Scott

ADVANCING TECHNIQUES TO DETERMINE RELATIVE SURFACE COVERAGE ON SMALL PARTICLES

It is necessary to know the relative surface coverage of molecular modifiers on gold nanoparticles for assessing their usefulness in clinical applications.

Engineered nanoparticles intended for use in clinical diagnostic and therapeutic applications require careful and comprehensive characterization in order to ensure the safety and efficacy of their use. Surface chemistry data, including chemical composition, chemical bonding, and susceptibility to oxidation are necessary to correctly interpret results from the testing of nanomaterials in biological matrices. In collaboration with scientists from the National Cancer Institute's Nanotechnology Characterization Laboratory, CSTL researchers are advancing measurement tools to evaluate the surface chemical composition of nanoparticle materials intended for clinical therapeutic or diagnostic applications. Measurement methods using x-ray photoelectron spectroscopy have been developed to investigate the surface chemistry, surface coverage, and stability (oxidation) of conjugate thiolated molecules on gold nanoparticles. The methods have been applied to gold nanoparticles with simple linear molecular modifiers and extended to more complex branched dendron modifiers. The chemical bonding environment between atoms of the molecular modifier and the gold atom surface can be studied with this technique. This technique will help advance the translation of engineered nanoparticles into the clinical realm.

Contact
Rebecca A. Zangmeister

DEVELOPING METHODS TO EXAMINE THE CHEMICAL COMPOSITION OF NANOMATERIALS

Nanoscale materials that are well characterized for chemical composition enable the quality and comparability of analytical results to be assessed.

Reference materials being developed by NIST for nanomaterials are focused on a number of physical and chemical parameters. A team of CSTL researchers are contributing to the progress of these materials by developing methods for the chemical compositional characterization of nanomaterials. At the nanoscale, single defects and slight changes to composition can dramatically influence reactivity. Proper characterization of the chemical composition is critical. Measurement capabilities being developed include the detection and quantification of nanoparticles by way of elemental signatures (such as gold and silver), quantification of other chemicals present in nanomaterials, and techniques to distinguish between free and complexed nanoparticles or to distinguish metals present as nanoparticles from other chemical forms of the metal. These efforts increase the usefulness of nanomaterial reference materials to users by providing information that may be relevant toward the usage and behavior of these materials in different measurement systems. Likewise, information regarding the chemical form of a nanomaterial, is critical in understanding their behavior for specific applications such as drug delivery. Determination of nanomaterial formulations being developed for cancer treatment are additionally underway in collaboration the National Cancer Institute's Nanotechnology Characterization Laboratory, further enabling the NCI to transfer potential life-saving treatments to clinical trials.

Contact
Gregory C. Turk
Lane C. Sander
Stephen A. Wise

GREEN APPROACHES FOR CARBON NANOTUBE PROCESSING

Methods for purification, size-separation and dispersion of carbon nanotubes are needed to prepare nanotubes for aqueous biological applications.

Since the discovery of the C60 "buckyball" potential applications of carbon nanotubes (CNTs) have become the focus of a broad range of academic and industrial research programs in areas ranging from microelectronics to clinical therapeutics. Current efforts at manipulating CNTs have typically focused on longer nanotubes (> 500 nm), so that materials studied were "nano" scaled in diameter only. While longer tubes are easier to manipulate with existing methods, these CNTs do not exhibit mechanical, electrical and biological properties of a true nanoscale material - *smaller is better*. While raw carbon nanotube materials can be generated relatively inexpensively, there still remains a lack of low cost, scalable purification techniques to produce high purity, homogenous CNT fractions. To address the bottleneck in post-production, scalable CNT processing, NIST researchers and collaborators at the University of Maryland Biotechnology Institute have developed a series of novel aqueous-based methods, methods that remove the use of toxic solvents. Using these methods, a step-wise enrichment and fractionation of commercially obtained CNT materials has been achieved. Protocols for uniformity in sample preparation and measurement have also been explored as these will be essential to enabling a quantitative comparison of low-to-high grade CNTs in their raw, intermediate and purified states. These will no doubt assist with the realization of the commercial and societal benefits CNTs have to offer.

Contact
John P. Marino

METHODS TO ASSESS NANOMATERIAL BIOINTERACTIONS

Methods are needed to understand the potential interactions of engineered nanoparticles with human DNA to better assess the safety of nanoparticles or products that contain them.

Nanotechnology research has resulted in the rapid creation of engineered nanomaterials with many foreseeable applications in medical imaging, diagnosis and in drug delivery. However, there is a notable scarcity of both acute and chronic human toxicity data for these new materials. Exposure to nanomaterials, is an immediate concern for human health and safety. This concern derives from the observation that certain nanomaterial characteristics, such as size, surface coating, and shape, have the capability to generate reactive oxygen species, reactive nitrogen species or other free radicals in biological systems. It is not known at a fundamental molecular level if engineered nanomaterials promote or hinder the formation of free radicals in the body or how these nanomaterials might interact with specific molecules such as deoxyribonucleic acid (DNA), the essential nucleic acid that contains the genetic instructions used in the development and functioning of the body. By using simple solutions of DNA and gold nanoparticles as a preliminary model system, CSTL researchers are generating fundamental data that will enable continued and deeper research into the potential mechanisms of nanoparticle interactions (if any) with the DNA in mammalian cells. The findings could benefit the nanomaterial market by making it possible to detect risks or benefits these materials may pose to human health or the environment.

Contact
Bryant C. Nelson
Miral M. Dizdar

MOVING CLOSER TO KNOWING THE EFFECT OF PARTICLE SIZE ON THE HEALTH OF CELLS

It is not clear whether many common toxicity protocols written for micrometer-size particles are applicable to nano-size based biomaterials and products.

Since the use of 1-100 nm nanoparticles may result in unknown biological risks and no clear confirmation exists to show that established standard test protocols for larger particles are appropriate, it is important to develop reliable methods to study the potential cytotoxicity and inflammatory response of nano-sized particles in vitro. In collaboration with the U.S. Food and Drug Administration and University of Maryland scientists, the biocompatibility of photoluminescent silicon nanoparticles (3.0 ± 0.1 nm) and silicon microparticles (100~3000 nm) are being tested with macrophage cells, a type of white blood that ingests (takes in) foreign material, from mice using standard protocols for micron sized particles to measure cytotoxicity and inflammatory responses. Photoluminescent silicon nanoparticles have a bright and stable fluorescence and are promising candidates for use in bio-imaging, cell staining and drug delivery. Fluorescence microscopy showed that the nanoparticles could enter macrophages, however because of their small size or atypical method of entry into the macrophage, nanoparticles are not recognized as "foreign" by the macrophages. Particle size was observed to be a decisive factor in determining the biological interactions of the silicone particles. The outcomes will help define testing and safety requirements for biomedical applications of nanoparticles.

Contact
Vytas Reipa

TRACKING NANOPARTICLES USING CHEMISTRY AND LIGHT

The unique chemical and physical properties of engineered nanomaterials that make them attractive for numerous applications also contribute to their unexpected behavior in the environment.

The potential environmental risks, including their impact on aquatic organisms, have been a central argument for regulating the growth of the nanotechnology sector. One area of significant concern for the risk assessment community is the potential for engineered nanomaterials to trophically transfer and accumulate (a process called biomagnification) along a particular food chain. Such behavior has been demonstrated for toxic contaminants such as DDT and mercury and necessitates human consumption limits of certain finfish, for example. A stable invertebrate community, consisting of the microbial loop and metazoan food web, are recognized as important components of productive aquatic ecosystems. Demonstration of such transfer would first require QD uptake from the liquid phase by the prey species (single celled organisms called ciliates) followed by ingestion of the ciliates, with subsequent assimilation of the QDs, by a predatory rotifer species. To that end, we utilized carboxylated and biotinylated QDs and a representative ciliate and rotifer species to investigate the trophodynamics of a model engineered nanomaterial in a simplified freshwater invertebrate food web. QDs were used in these experiments due to their straightforward detection with both optical and transmission electron microscopy, their unique elemental composition that allows nanomaterial quantification, and their variety of available surface functionalities. The findings indicate that dietary uptake of nanomaterials should be considered for higher trophic aquatic organisms. However, limited bioconcentration and lack of biomagnification may impede the detection of nanomaterials in invertebrate species. Such data are necessary to assess if and how engineered nanomaterial move in the environment.

Contact
R. David Holbrook

MEASUREMENTS & STANDARDS
Activities Beyond Our Borders

CSTL engages in a number of international activities to support trade and global science, and to promote the international acceptance of U.S. measurement standards. Long-term involvement and collaborations with National Metrology Institutes around the world have become more formal in recent years with the signing of the European Union and United States Mutual Recognition Agreement and subsequent implementation arrangements with the International Committee of Weights and Measures. CSTL domestic and international standards efforts are critical to U.S. industry by assuring comparability of measurements across borders, thus facilitating fair trade and increasing our nation's competitiveness in the global market place.

USING SOUND TO M

Acoustic Thermometry

By measuring the speed of sound in argon gas, CSTL researchers have determined the thermodynamic temperature of the gas with unprecedented accuracy in the range 0 °C to 552 °C. The findings are used to improve the accuracy of platinum resistance thermometry, resolve discrepancies in the benchmark values upon which the International Temperature Scale of 1990 is based, and provide data upon which the next temperature scale will be partially based. Data from the NIST acoustic thermometer will form an improved basis for the equations used with platinum resistance thermometry, which is the primary mechanism to disseminate accurate temperature standards.

Thermodynamic temperatures in the industrial processing environment are difficult to measure, and new techniques are continually being developed to make the mathematical formulas in the International Temperature Scale of 1990 (ITS-90) more closely represent the laws of nature. Recent innovations by researchers in CSTL are among the advances working toward new and improved reference thermometers for industry and research.

Acoustic thermometry, which involves measuring the speed of sound waves in a basketball-sized acoustic resonator filled with gas, and then using this value and fundamental physical properties to calculate temperature. The thermometer has several novel features including:

- Continuous gas purging of the resonator to minimize gas impurities,

- A gas-chromatography system to measure the impurities in the gas exiting the resonator,

- Measurements on the ITS-90 with up to five long-stem standard platinum resistance thermometers.

Recent publications report results up to 552 K. The goal of reducing the uncertainty of thermodynamic temperature by a factor of five in the range 273 K to 552 K has been achieved. These data are in excellent agreement with radiometric measurements (at 700 K and above), other acoustic measurements (at 390 K and below), and one of the past gas thermometry results. In 2008, techniques were developed to perform precision acoustic measurements with a microphone at room temperature. The technique has pushed the upper limit of CSTL's data to 650 K.

Contact
Dean Ripple

Johnson Noise Thermometry

A primary temperature measurement technique based on the fundamental properties of thermal fluctuations in conductors is useful for both temperature scale metrology and for the development of reliable thermometers for harsh environments. CSTL researchers measure these fluctuations with respect to quantum electronic noise sources and-or other thermal noise sources. The NIST-developed spectral Johnson Noise Thermometry (JNT) methods link the measured thermal voltage directly to the as-maintained electrical units via the AC Josephson Voltage Standard. This new methodology conveys key advantages from a metrology standpoint and will extend the practical application range for JNT systems.

Johnson noise thermometry is essentially a system that relies on the "noise" of jiggling electrons as a basis for measuring temperatures with extreme precision. The system is nearly precise enough now to help update some of the crucial underpinnings of science, including the 54-year-old definition of the Kelvin, the international unit of temperature. The JNT system developed by CSTL and other NIST researchers represents a five-fold advance in the state of the art in noise thermometry thanks to its use of a unique quantum voltage source combined with recent reductions in systematic errors and uncertainty. It is also simpler and more compact than other leading systems for measuring high temperatures, such as those based on the pressure and volume of gases.

CSTL researchers apply new technologies and methodologies to JNT using wide-band spectral analysis, digital signal processing, and pulse-quantized voltage synthesis. The findings are useful for verification of the International Temperature Scale of 1990 over a temperature range from 500 K to 930 K. This range overlaps those covered by other thermodynamic methods, most notably the upper-most range of Acoustic Thermometry and the lower-most range of absolute Radiation Thermometry. The JNT results provide independent data for scale validation in these overlapping ranges. Future impacts in advancing the application of JNT temperature measurements in new environments, such as high-temperature nuclear reactors, are expected.

Contact
Weston L. Tew

INTERNATIONAL FLUID PROPERTY STANDARDS: NIST REFPROP

CSTL researchers have released an expanded version of a popular NIST database, a computer package for calculating the properties and modeling the behavior of fluids. Data on key components of alternative fuels, such as ethanol and hydrogen, are among the many new additions to NIST's Reference Fluid Thermodynamic and Transport Properties Database Version 8.0, which provides critically evaluated property values needed to evaluate fluids and optimize related equipment and processes. The standard reference data provided by REFPROP allow international consensus on information for trade and technology.

Major improvements have been made to NIST's Reference Fluid Thermodynamic and Transport Properties Database Version 8.0, which provides critically evaluated property values needed to evaluate fluids and optimize related equipment and processes. This new update adds new standards that either have been recently adopted, or are in the process of becoming standards. Most notably are the AGA-8 and GERG-2004 equations for natural gas.

AGA-8 has been the current standard in the U.S. for natural gas properties since 1992, but was not available in version 7.0 of REFPROP since it was only valid in the gas phase. This standard did not have the ability to calculate phase equilibrium properties, including dew points and bubble points. Although properties at the dew point were within the limits of the AGA-8 equation, the unknown equilibrium point became a real hindrance for natural gas processors. The European GERG-2004 equation set out to overcome this deficiency and to add properties in the liquid phase and critical region. It was designed to be more accurate than the AGA-8 equation in the gas phase, so that it could become a new standard for the gas industry. This equation is currently being adopted by ISO. The new equation improves the representation of natural gas properties, properties of mixtures for other components that are cryogenic or contain carbon dioxide.

NIST has distributed thousands of copies of REFPROP since 1990, and thousands of additional copies are made available every year through commercial third party distributors. REFPROP has been the *de facto* standard in the refrigeration industry for years and is credited with helping industry find replacements for ozone-depleting chlorofluorocarbons (CFCs).

Contact
Eric W. Lemmon

MEASUREMENTS, STANDARDS, AND REFERENCE MATERIALS FOR INDUSTRIAL COMMODITIES

All industrial sectors rely on elemental analysis and physical properties testing to confirm product compliance with manufacturing specifications. Support of product compliance testing has long been a core activity in CSTL in collaboration with industry associations and standards developing organizations (SDOs). For years CSTL has invested heavily in development of standard test methods for industrial commodities and in SRM development to validate those methods and to establish traceability to the International System of Units (SI). These efforts are aligned with industry needs through interaction with SDOs, industry associations, expert private sector laboratories, and commercial reference materials producers.

In the arena of industrial commodities, the primary challenges are to provide measurement tools and reference materials that allow U.S. industry to establish comparability of measurement results to results obtained by customers, competitors, and regulators for the basic chemical properties of products, intermediates, and by-products on a worldwide basis. All members of a given industrial supply chain benefit from critically evaluated standard test methods validated using trusted reference materials.

The private sector can leverage NIST reference materials for key commodities by creating their own reference materials targeted at products having more specialized compositions and applications.

The following SRMs have been developed or renewed in collaboration with industry:

- SRMs for steel, Portland cement, polyethylene, silicon metal, zirconium, and copper mine tailings,

- Upgraded existing SRM certificates to comply with ISO Guide 31 for more than 20 ferrous alloys, nonferrous alloys, and geological materials,

- Initiated SRM development projects for free-cutting brass, lead-free solder, molybdenum concentrates, silicon carbide, copper ore, refined copper, and feldspar.

These efforts increase the availability of reference materials for industrial commodities including metals, ores, cement, polymers, glass, and more. Activities include the development of international standard test methods for elements in titanium, elements in plastics, and compounds in cement.

Contact
John R. Sieber
Gregory C. Turk
Stephen A. Wise

A DYNAMIC EXPERT SYSTEM FOR THERMODYNAMIC DATA

The ThermoData Engine (TDE) – NIST Standard Reference Database 103 - incorporates all major stages of a dynamic expert system including data retrieval, grouping, normalization, sorting, consistency enforcement, fitting, and prediction. The third version of TDE (now NIST SRD 103b), released in 2008 by CSTL researchers, expanded the implementation of the concept to the thermophysical properties of binary mixtures. TDE provides access to single-phase thermodynamic and transport property data for more than 30,000 mixtures and performs automated evaluation of most of those properties. Proprietary data can be entered for inclusion in the evaluation, and the user can influence the evaluation process by changing relative data weights or by rejecting particular data sets.

Traditionally, critical data evaluation is an extremely time- and resource-consuming process, which includes extensive use of labor in data collection, data mining, analysis, fitting, etc. Because of this, it must be performed far in advance of a need within an industrial or scientific application. In addition, it is quite common that by the time the critical data-evaluation process for a particular chemical system or property group is complete (sometimes after years of data evaluation projects involving highly skilled data experts), it must be re-initiated because significant new data have become available. This type of slow and inflexible critical data evaluation can be defined as 'static.'

To address the weaknesses of 'static" evaluations, the concept of a dynamic data evaluation system has been developed by CSTL researchers. This concept requires large electronic databases capable of storing essentially all experimental data known to date with detailed descriptions of relevant metadata and uncertainties. The combination of these electronic databases with expert-system software, designed to automatically generate recommended data based on available experimental data, leads to the ability to produce critically evaluated data dynamically or 'to order'. This concept contrasts sharply with static critical data evaluation, which must be initiated far in advance of a particular need. The dynamic data evaluation process dramatically reduces the effort and costs associated with anticipating future needs and keeping static evaluations current.

Contact
Michael Frenkel

DATA COMMUNICATION FROM GENERATION TO ULTIMATE USE

ThermoML is an Extensible Markup Language (XML)-based new International Union of Pure and Applied Chemistry (IUPAC) standard for storage and exchange of experimental, predicted, and critically evaluated thermophysical and thermochemical property data. ThermoML has been developed to serve as standardizing platform-independent interoperable tool for delivery of thermophysical and thermochemical property data from "data producers" to "data users" as well as for information exchange between experimentalists, journals, organizations, products, and users. Since 2002, the NIST Thermodynamics Research Center within CSTL has played a key role in the development and establishment of the international standard.

Thermodynamic property data represent a key resource for development and improvement of all chemical process technologies. Rapid growth in the number of custom-designed software tools for engineering applications has created an interoperability problem between the formats and structures of thermodynamic data files and required input/output structures for the software applications. Establishment of efficient means for communication of thermodynamic data is absolutely critical for provision of solutions to such technological challenges as elimination of data processing redundancies and data collection process duplication, and rapid data propagation from measurement to data management system.

Since 2002, the NIST Thermodynamics Research Center within CSTL has played a key role in leading the efforts of the Task Group for the IUPAC project "XML-based IUPAC Standard for Experimental and Critically Evaluated Thermodynamic Property Data Storage and Capture". ThermoML is an Extensible Markup Language (XML)-based new IUPAC standard for storage and exchange of experimental, predicted, and critically evaluated thermophysical and thermochemical property data. In 2007 IUPAC approved a proposal submitted by NIST in cooperation with other industrial and academic institutions to further expand ThermoML to provide capabilities to communicate thermodynamic properties of biomaterials as well as properties of electrolytes including speciation issues. The next version of the ThermoML is scheduled for release in 2009 and will address thermodynamic properties of biomaterials.

Contact
Michael Frenkel

INTERNATIONAL STANDARDS ON THE PROPERTIES OF WATER

Water is one of the most important industrial fluids and it is important to have internationally accepted standards for its properties. This information is crucial to the analysis and design of thermal power cycles. The International Association for the Properties of Water and Steam (IAPWS) is an international non-profit association of national organizations concerned with the properties of water and steam, particularly thermophysical properties and other aspects of high-temperature steam, water and aqueous mixtures that are relevant to thermal power cycles and other industrial applications. CSTL researchers are active in the leadership and working groups of IAPWS and have led the development of several recent standards adopted by IAPWS.

The properties of water are key to many technologies, and international standards are required to ensure global acceptance of designs, equipment, and instrumentation. CSTL researchers play a leading role in the development of several new international standards for the properties of water. This work was performed under the auspices of the International Association for the Properties of Water and Steam (IAPWS). Contributions included the development of a new reference formulation for the viscosity of water, new "backward equations" for properties which substantially decrease the computation time for steam-turbine calculations, a new seawater standard for oceanographic and engineering applications, and simple but highly accurate correlations for thermodynamic properties, viscosity, thermal conductivity, and static dielectric constant of liquid water at atmospheric pressure.

The following are associated products:

- IAPWS Formulation 2008 for the Thermodynamic Properties of Seawater
- IAPWS Formulation 2008 for the Viscosity of Ordinary Water Substance
- Supplementary Release on Properties of Liquid Water at 0.1 MPa
- NIST/ASME Steam Properties (NIST SRD 10), Version 2.22, 2009
- NIST Reference Fluid Thermodynamic and Transport Properties Database (NIST SRD 23, REFPROP), version 8.1, 2009
- NIST Chemistry WebBook

Contact
Allan H. Harvey
Marcia L. Huber

MATERIALS FOR ENVIRONMENTAL MONITORING

For the past 40 years NIST has developed SRMs for the determination of inorganic and organic contaminants in environmental matrices such as sediments/soils, marine and animal tissues, air particulate, and botanical materials. These natural environmental-matrix SRMs for contaminants are used worldwide as the basis for validating accuracy and comparability within the environmental measurement community. For inorganic analysis, the natural matrix SRMs typically have concentration values assigned for toxic metals and other elements of interest. Typical organic contaminants with values assigned in these natural matrix SRMs include polycyclic aromatic hydrocarbons (PAHs), polychlorinated biphenyls (PCBs), and persistent chlorinated pesticides. Recent activities have focused on expanding the number of PAHs and PCB congeners with values assigned and on assigning values for new classes of compounds such as brominated flame retardants.

In 1971 NIST issued the first natural environmental matrix SRM for contaminants, SRM 1571 Orchard Leaves, with certified concentrations for trace elements. During the 1970s, additional natural matrix SRMs for trace element content were developed using this approach including bovine liver, fly ash, spinach, pine needles, water, river and estuarine sediment, air particulate matter, and oyster tissue.

A decade later the first environmental matrix SRM for organic contaminants was issued, SRM 1649 Urban Dust/Organics, with certified concentrations for a limited number PAHs. SRM 1649 and subsequent natural matrix materials issued during the next decade (coal tar, diesel particulate matter, marine sediment, and mussel tissue) established the multiple methods approach for organic contaminants in environmental matrices for PAHs, polychlorinated biphenyls, and chlorinated pesticides. During the past four decades, NIST has issued over 60 natural environmental matrix SRMs with certified values for inorganic and/or organic contaminants. Many of these SRMs have been developed specifically to address the regulations and needs of the U.S. Environmental Protection Agency and the National Oceanic and Atmospheric Administration as part of their air and ocean water quality monitoring programs in the United States.

Contact
Stephen A. Wise
Michele M. Schantz
Gregory C. Turk

RELATIVE QUANTIFICATION OF GENOMIC DNA FRAGMENTS: CCQM BAWG-P113

The provision of a traceable standard to the biological community is an area of active research for many National Measurement Institutes (NMIs). The quantification of the relative amount of DNA sequences extracted from a biological tissue remains a complex analytical procedure and relies on the availability of such standards. CSTL researchers participated this year in a study to demonstrate the ability to quantify DNA sequences present in a biological tissue using an independent calibrant system, in this case, using a plasmid calibrant for genomic DNA. This study will help determine if the calibrant system is fit for the intended use of quantifying the amount of biotech material in a batch of maize.

In the past year, a study was organized by the Institute for Reference Materials and Measurement (IRMM), Geel, Belgium to demonstrate the comparability of results from the quantification of DNA sequences present in a biological tissue. Fourteen NMIs participated and results were presented at the Consultative Committee for Amount of Substance (CCQM), Bioanalytical Working Group (BAWG) meeting in November 2008.

Participants were provided with a material to serve as a calibrant for quantitative PCR consisting of a plasmid containing the DNA sequences of segments of an endogenous gene target and a transgenic gene target. The endogenous gene target came a maize specific target. The transgenic target came from a genetically engineered maize event named 1507. The unknown samples to be analyzed were maize powders that contained a defined mass fraction of genetically engineered maize 1507 mixed with wild type maize. The unknown samples were certified reference materials which had already been tested for homogeneity and stability. Results are published in the Draft Report of the CCQM-P113 (Oct 31, 2008) and CSTL results were well within the values calculated by the NMI hosting the pilot study. Based on these preliminary data, use of a plasmid calibrant for relative quantification of DNA sequences extracted from a biological tissue may be fit for the intended use of quantifying the amount of genetically engineered material in a batch of maize. In addition, use of this calibrant may permit correct labeling of food products.

Contact
Marcia J. Holden

YEAR IN REVIEW

Year in Review

Alejandro Herrero, director of the IRMM, signed the agreement in Geel, Belgium, on behalf of the EC, and Willie May, director of NIST's Chemical Science and Technology Laboratory, signed for the United States.

NIST, EC Agency Partner for Better Measurements and Standards

Enhancing trade between the United States and the nations of the European Union (EU) while helping ensure the safety and quality of goods sold in both markets is the goal of a collaborative agreement signed on Dec. 17, 2007, between the European Commission (EC) Joint Research Centre (JRC)'s Institute for Reference Materials and Measurements (IRMM) and NIST. The pact will advance the development and availability of international measurement standards in the fields of chemistry, life sciences and emerging technologies.

Under the agreement, the JRC and NIST will work to better coordinate their research and development programs in metrology. This will include collaborative research on new measurement methods and their quality assurance, including but not limited to cooperation in the preparation and value-assignment of certified reference materials. The JRC and NIST also plan to share resources and harmonize their respective regional and national responsibilities for chemical metrology, biometrology and international measurement standards.

Membrane Model May Unlock Secrets of Early-Stage Alzheimer's

CSTL researchers and three collaborating institutions are using a new laboratory model of the membrane surrounding neurons in the brain to study how a protein long suspected of a role in early-stage Alzheimer's disease actually impairs a neuron's structure and function. Studying the beginnings of Alzheimer's is nearly impossible in humans because by the time the disease is diagnosed, most patients have moved into its later stages.

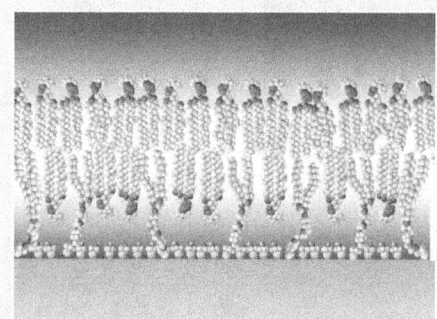

Diagram of NIST's "tethered bilayer membrane" model shows the silica surface covered with gold at the bottom.

The team of researchers have developed a laboratory model that recreates a simplified version of the nerve cell membrane, allowing the study of Alzheimer's disease mechanisms at the molecular level. A clever piece of molecular-level design, the system is built by first covering a silica surface with gold. Sulfur atoms, which bond well to gold, are then added to act as anchors to hold the bilayer membrane. The result is a stable, tethered membrane with an aqueous environment on both sides that accurately models the behavior of the nerve cell membrane.

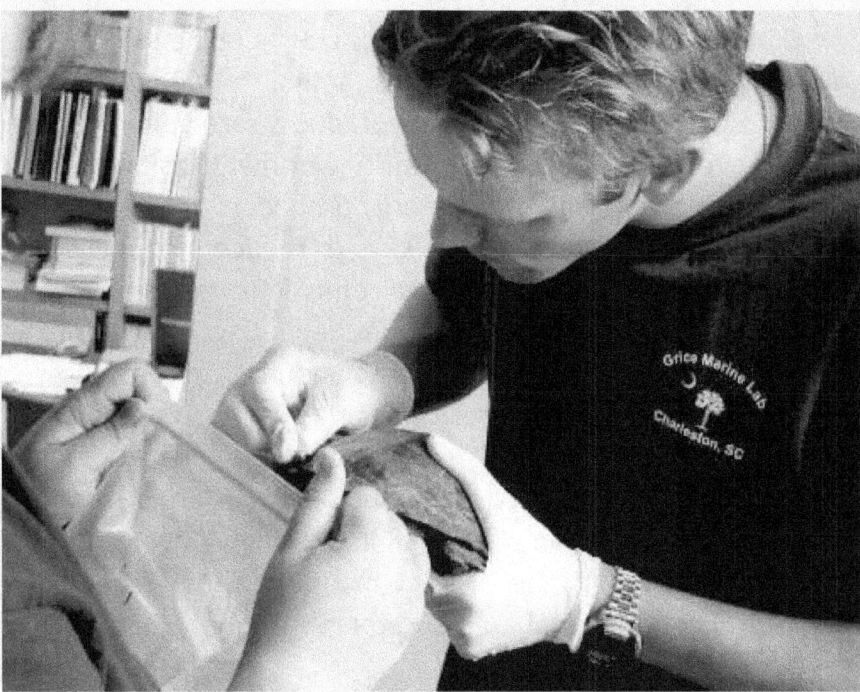

College of Charleston graduate student Jeff Schwenter takes samples from turtles. Researchers here and at the Hollings Marine Laboratory are studying mercury levels in turtles.

Turtles Shell Out Information on Mercury

Biologists in Charleston have discovered that turtles are reliable detectors of local mercury contamination, a finding that could have major implications in how the toxin is measured and regulated. CSTL researchers at the Hollings Marine Laboratory at Fort Johnson are taking blood samples and shell scrapings from turtles along the Eastern seaboard and testing these samples for mercury. These studies are being conducted in collaboration with biologists from the College of Charleston, the Medical University of South Carolina, and the South Carolina Department of Natural Resources.

So far, they've found that sea turtles captured near the mouths of rivers had higher mercury levels than those caught offshore. In estuaries they discovered that turtles near coal-fired power plants and other industrial sites also had elevated mercury levels.

Mercury tends to build up in turtles' shells over time, just as it does in human hair. But until recently, scientists weren't sure whether measuring these would be reliable indicators of mercury in turtle tissue. "There just wasn't a lot of data on mercury," said Rusty Day, one of the CSTL biologists.

Biologist Russell (Rusty) Day and other researchers have tested turtles for mercury and are finding that they make good indicators for whether an area has a local mercury contamination problem.

Using turtles to monitor mercury levels could add a new twist to the high-stakes debate over whether coal-fired power plants and other industries create local mercury pollution problems. "Once people link local mercury contamination to local mercury-spewing plants, pressure will increase to clean those plants up, and that costs money," said Blan Holman, a lawyer for the Southern Environmental Law Center.

International Biofuels Effort Seeks Fewer Barriers, More Trade

On February 1, 2008, the governments of the United States, Brazil and the European Union (EU)—the world's major producers of biofuels—released an analysis of current biofuel specifications with the goal of facilitating expanded trade of these renewable energy sources.

Spurred by increased market demands, this report was solicited by the U.S. and Brazilian governments and the European Commission (EC) on behalf of the EU, with the work conducted by an international group of fuel standards experts. Biofuels-derived from biological materials such as plants, plant oils, animal fat and microbial byproducts—are gaining popularity worldwide as both energy producers and users seek ways to reduce greenhouse gas emissions, move away from dependence on fossil fuels and invigorate economies through increased use of agricultural products. As a result, biofuels are becoming an increasingly important commodity in the global marketplace.

Biofuels can originate from many sources such as soy, palm oil, and animal fat. Creating reference materials to support development and testing of biofuels, and analytical measurement methods for source identification are part of CSTL collaborations with Brazil's National Institute of Metrology.

Recognizing that many of the issues relating to variations in specifications can be traced to different measurement procedures and methods, CSTL researchers and Brazil's National Institute of Metrology, Standardization and Industrial Quality (Instituto Nacional de Metrologia, Normalização e Qualidade Industrial or INMETRO)—are collaborating on the development of joint measurement standards for bioethanol and biodiesel to complement the efforts of standards developing organizations. Brazil, the world's biggest exporter of ethanol, already requires up to a 25 percent blend of ethanol with all gasoline that is sold. In the United States, the Energy Policy Act of 2005 sets a 7.5 billion gallon goal for national biofuel consumption (usually ethanol) by 2012.

A Molecular 'Salve' to Sooth Surface Stresses

CSTL researchers have shown for the first time that a single layer of molecular "salve" can significantly soothe the stresses affecting clean metal surfaces. In addition to confirming that the application of a monolayer did reduce surface stresses, the team also discovered that the

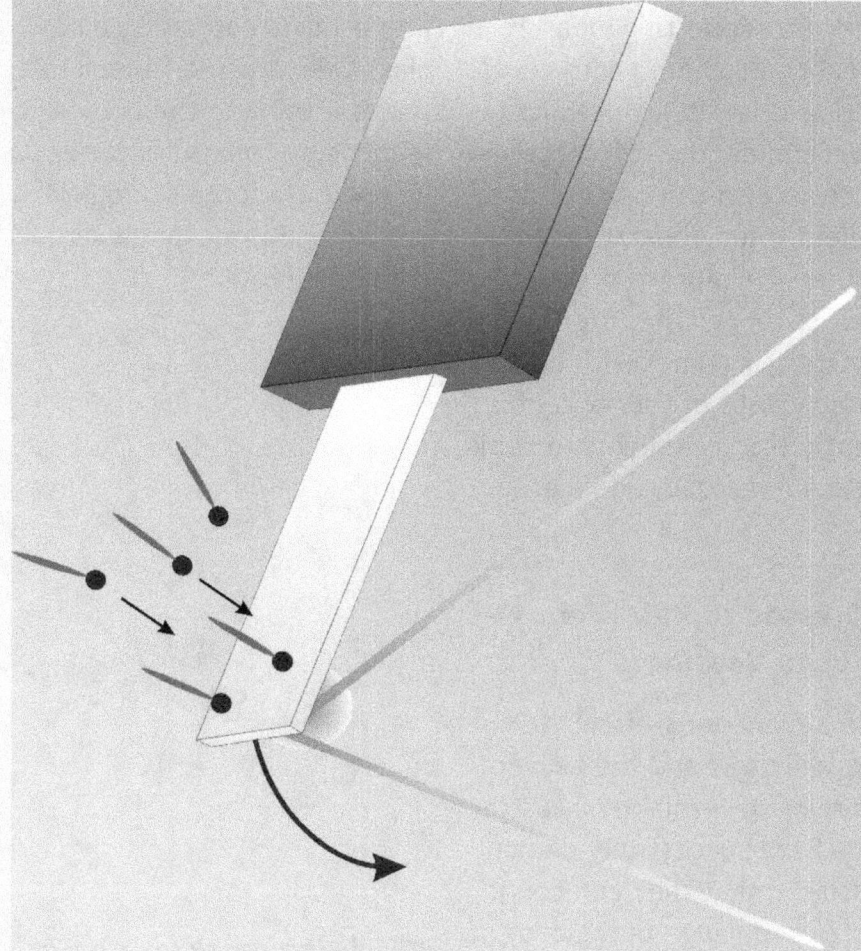

CSTL researchers measured the surface stress of a monolayer film on gold by measuring the changing curvature of a gold-coated glass cantilever as molecules of mercaptobenzoic acid were deposited on the gold. The change in curvature was detected by the shifting reflection of a laser beam impinging onto the back side of the glass.

monolayer would provide a proportional response to the amount of the substance it was designed to detect, which would result in a quantifiable decrease in the tension of the cantilever.

Lead Poisoning Detection Improved

Lead in goat blood might not be on the top of your shopping list, but for U.S. medical personnel who each year perform more than 2 million human blood measurements, Standard Reference Material (SRM) 955c, developed by CSTL researchers, can't be beat. SRM 955c is an improved version of SRM 955b, a material clinicians already relied on heavily to provide quality assurance for lead in blood measurements. Significant changes in material composition, lead levels and expanded uncertainties of the certified lead concentrations

longer the molecules were allowed to sit the more comfortable they became with their new surroundings. As the monolayer became more comfortable, it became more stable, and the atoms in the metal began to adopt the molecules into the family, which substantially reduced the surface stresses. The findings provide a deeper understanding of the forces at work at the interface of molecules and surfaces. Most notably the discovery could be used to create a new generation of chemical and biological sensors. These sensors would use molecular monolayers deposited on metal surfaces that are manufactured to react in the presence of chemical or biological agents in the environment. The activation of the

Deteriorating paint on window sills can get on the hands of children and be ingested, increasing the likelihood of lead poisoning if it's an older home.

make SRM 955c an even more effective tool for use in addressing lead poisoning, a condition particularly harmful to the developing nervous systems of fetuses and young children, causing learning disabilities and behavior problems and, at high levels, seizures, coma and death.

Determining the Quality of Ginkgo

CSTL researchers have issued a suite of Standard Reference Materials (SRMs) for ginkgo biloba, one of the most popular dietary supplements in the marketplace, with annual worldwide sales estimated at $1 billion. The reference materials are designed to help researchers validate the accuracy of analytical methods for flavonoids and terpene lactones (plant constituents that may be associated with the perceived effectiveness of ginkgo) as well as toxic elements in ginkgo. In addition to supporting measurements associated with clinical trials or verifying product label claims, the reference materials also can be used by dietary supplement manufacturers to improve product consistency. The reference materials comes with certified values of four potentially toxic trace elements: arsenic, cadmium, lead, and mercury.

Ginkgo leaves.

Dressed to Kill: From Virus to Vaccine

In a pioneering effort, CSTL researchers and the University of Queensland in Australia have successfully demonstrated that they can count, size and gauge the quality of virus-like particle-based (VLP) vaccines much more quickly and accurately than previously possible. Their findings could reduce the time it takes to produce a vaccine from months to weeks, allowing a much more agile and effective response to potential outbreaks.

"It takes a long time to develop vaccines because viruses have to be grown in chicken eggs or cell culture, which can take months," said Leonard Pease, a CSTL researcher working on the project. In order to speed the creation and delivery of these life-saving treatments, a new class VLPs dressed to look like the real thing to the body's immune system is being developed. It contains no DNA or RNA and is incapable of causing infection.

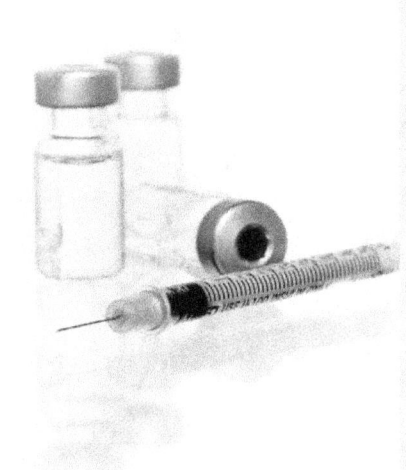

Virus-like particles do not have to be grown. As a result, vaccines based on these particles can be deployed much faster than traditional vaccines.

Whether or not a VLP-based vaccine will be effective depends on whether the VLPs are well-formed and properly coated. Electrospray differential mobility analysis is a particle sizing technique able to count millions of particles an hour with subnanometer resolution. NIST researcher Leonard Pease and his team were able to size well-formed VLPs that have been coated with bird flu proteins, a critical step towards the creation of future bird flu vaccines.

NIST Receives Citation for Chemical Breakthroughs

From the American Chemical Society Division of the History of Chemistry

NIST and Columbia University have been named by the American Chemical Society (ACS) Division of the History of Chemistry as recipients of the ACS Citation for Chemical Breakthrough Award in recognition of the isolation of deuterium, the isotope of hydrogen commonly called "heavy hydrogen." The feat—proving that deuterium actually existed—was achieved in 1931 by physicists from the National Bureau of Standards (NBS), NIST's predecessor, and Columbia.

In 1931, NBS physicist Ferdinand Brickwedde collaborated with Columbia physicists Harold Urey and George Murphy to produce the first sample of deuterium at the NBS Low Temperature Laboratory in Washington, D.C. (on what is now the campus of the University of the District of Columbia). Team leader Urey was awarded the Nobel Prize in Chemistry for the discovery. Today, deuterium is used as a non-radioactive tracer in scientific and medical research and in the study of thermonuclear fusion reactions.

The ^{2}H isotope, as early as in 1934, was named by Columbia physicist Harold Urey as deuterium (symbol D) with a nucleus called deuteron (symbol d).

NIST, UMBI To Expand Cooperation in Bioresearch

In August 2007, officials from NIST and the University of Maryland Biotechnology Institute (UMBI) signed a Memorandum of Understanding designed to expand the scope of joint research and educational activities in the biosciences between the two institutions. The new MOU updates an existing relationship between NIST's CSTL and the University of Maryland that dates back to 1985, when they joined with Montgomery County, Md., to establish the Center for Advanced Research in Biotechnology (CARB), a joint research venture emphasizing work on the relationship between structure and function in biomolecules and the development of new technologies for the measurement, analysis and design of biomolecules. The MOU, which provides a framework for future joint activities, allows for interdisciplinary research programs that leverage NIST's measurement and analysis expertise across the range of physical sciences with UMBI's resources and allows for increased exchange of staff via temporary appointments and educational opportunities.

UMBI Center for Advanced Research in Biotechnology in Shady Grove, MD.

Department of Commerce (DoC) Honorary Awards

*The **DoC Gold Award** is given for distinguished performance characterized by extraordinary, notable, or prestigious contributions that impact the mission of the Department.*

David Duewer, John Butler, Peter Vallone, Janette Redman, and Margaret Kline were recognized for their research and measurement services in the area of human identity testing to support the criminal justice, legal, and disaster recovery communities.

---

*The **DoC Silver Award** is given for exceptional performance characterized by noteworthy or superlative contributions which have a direct and lasting impact within the Department.*

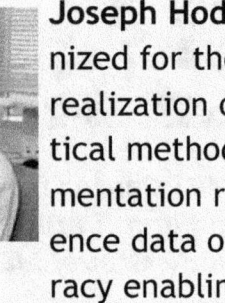

Joseph Hodges was recognized for the conception and realization of innovative optical methods and instrumentation resulting in reference data of superior accuracy enabling next generation trace gas analysis.

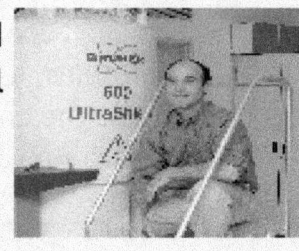

John Marino was recognized for his development of novel approaches to determine RNA structures and interactions based on NMR and fluorescence measurements.

Leonard Pease, Michael Winchester, and Rebecca Zangmeister and other members of the Gold Nanoparticle Team were recognized for developing and producing the first nanoparticle reference materials in the 10 nm to 60 nm size range.

Anne Plant was recognized for her leadership of NIST's program in Quantitative Cell Biology. She and her team have documented performance characteristics of their measurement approach and disseminated those concepts and the results of her measurements to the broader user community.

Stephen Wise was recognized for his outstanding achievement in organic chemical measurement science and Standard Reference Material development.

Department of Commerce (DoC) Honorary Awards

The **DoC Bronze Award** is given superior performance characterized by outstanding or significant contributions which have increased the efficiency and effectiveness of NIST.

John Elliott was recognized for his outstanding efforts regarding the development of methods for bioimaging and quantitative analysis of complex cellular characteristics.

Robert Greenberg was recognized for his outstanding leadership and contributions to the establishment of neutron activation analysis as a primary method of chemical analysis.

Peter Linstrom was recognized for his outstanding efforts leading to the development of the NIST Chemistry WebBook disseminated those concepts and the results of her measurements to the broader user community.

John Henry Scott was recognized for his exceptional innovation in the development and application of measurement methods that are enabling the dimensional and chemical compositional characterization of materials at the nanoscale.

Dmitri Tchekovskoi was recognized for his outstanding efforts regarding the development of the "International Chemical Identifier", a robust, extensible, and precise method for identifying molecular species.

Weston Tew was recognized for his leadership in developing new state-of-the-art methods in Johnson Noise Thermometry, placing NIST at the forefront of the international efforts in this field.

James Whetstone and the USMS Project Team were recognized for their assessment of the U.S. Measurement System based on the analysis of measurement-related needs for supporting innovation across 11 industrial sectors.

NIST-Named Incentive Awards

The NIST-Named Incentive Awards provides NIST the opportunity to recognize and publicize the valuable contributions of the NIST staff.

Mary Satterfield and other members of the NIST 2006 Diversity Day Committee received the *2007 NIST Diversity Award* for their significant contributions toward promoting a healthy, equitable, and diverse work environment at NIST.

Robert Chirico, Michael Frenkel, Andrei Kazakov, and Chris Muzny, received the *2008 Judson C. French Award* for their design, development, and implementation of the Web Thermo Tables (WTT), the first NIST Web Standard Subscription Database.

Arno Laesecke received the *2007 NIST Boulder Building Tomorrow's Workplace Award* for his excellence in working with students at all levels, such as the SURF and PREP student programs.

External Organizational Awards

Donald Archer received the *ASTM Achievement Award* for his work on gaseous hydrogen standards. The ASTM Achievement Award is presented to individuals who have made an outstanding contribution.

John Butler received the *2008 Arthur S. Flemming Award*, established by the Washington DC Jaycees in 1948. The Flemming Awards honor outstanding federal employees.

Robert Greenberg received the *2007 Hevesy Medal Award*. The Hevesy Medal Award is the premier international award of excellence to honor outstanding achievements in radioanalytical and nuclear chemistry.

Michael Kurylo received the *2008 U.S. Environmental Protection Agency Ozone Layer Protection Award* and will receive the 2009 Hillebrand Prize Award.

Eric Lemmon received the 2008 Natural Gas Sampling Technology Conference Distinguished Presentation Award for "The GERG-2004 Equation of State: A Wide Range Reference for Natural Gases."

Laurie Locascio received the *Arthur F. Findeis Award for Achievements by a Young Analytical Scientist* from the American Chemical Society, Division of Analytical Chemistry. The Findeis Award is given to recognize outstanding contributions to the field of analytical chemistry by a young analytical scientist.

William MacCrehan received the *2007 Distinguished Service Award, ASTM E54, Homeland Security Applications Award* for the development of E2520 "Standard Practice for Verifying the Minimum Acceptable Performance of Trace Explosives Detectors." The ASTM Achievement Award is presented to individuals who have made an outstanding contribution.

Ray Radebaugh received the *Cryogenic Engineering Conference Samuel C. Collins Award* for his outstanding contributions to the identification and solution of cryogenic engineering problems.

Katherine Sharpless received the *2007 AOAC Reference Material Achievement Award as a Fellow of AOAC International*. AOAC recognizes contributions of the technical division in the pursuit of the goals and objectives of AOAC International.

Elections & Nominations

Donald Burgess was appointed new *Technical Co-Editor* of the Journal of Physical and Chemical Reference Data. The Journal is published by the American Institute of Physics for NIST. The objective of the Journal is to provide critically evaluated physical and chemical property data, fully documented as to the original sources and the criteria used for evaluation.

David Duewer was appointed to a 3-year term on the *News and Features Advisory Panel of the Analytical Chemistry Journal*. The Journal explores the latest concepts in analytical measurements and the best new ways to increase accuracy, selectivity, sensitivity, and reproducibility including the latest peer-reviewed research and select significant applications.

Allan Harvey was appointed new *Technical Co-Editor* of the Journal of Physical and Chemical Reference Data. The objective of the Journal is to provide critically evaluated physical and chemical property data, fully documented as to the original sources and the criteria used for evaluation.

Hratch Semerjian was elected to the *NIST Gallery of Distinguished Alumni*. This honor is bestowed upon prominent NIST Alumni. Dr. Semerjian is a former Deputy Director of NIST and NIST Acting Director. He also served as the Director of CSTL for eleven years.

Steve Stein was elected a *Member of the Board of Directors for the American Society of Mass Spectrometry*. The Society is a professional society for those employed or working toward a degree in an area related to the field of mass spectrometry.

Distinguished CSTL Guests & Speakers

Staff from **Abbott Laboratories** in California and Texas, visited NIST in June 2008 and completed various laboratory tours to discuss NIST's activities in the Biosciences and Bioengineering.

Montgomery County Executive Isiah Leggett visited NIST in July 2008 to learn more about NIST's scientific leadership for the Nation's measurement and standards infrastructure and NIST's activities in the physical, chemical, and biological sciences. Exec. Leggett toured various labs at NIST and visited CSTL's joint institute, the Center for Advanced Research in Biotechnology (CARB, see page 99 for a description of CARB).

Staff from the **National Institute for Biological Standards and Control, UK**, visited with CSTL and NIST to participate in a workshop to address the need for standards in the Biosciences in October 2008.

Faculty members from the four **SEA-participating Historically Black Colleges** and Universities and other under-represented Universities, traveled to NIST to explore potential collaborations and partnerships. These included the University of North Carolina, Pembroke, the University of Maryland Eastern Shore, Hampton University, Prairie View A&M, Texas Southern University, Caflin College, Alabama A&M, Howard University, and representatives from the National Hispanic Universities.

Representatives from various **Biotechnology Companies in Spain**, including Histocell, S.I., Noray Bioinformatics, S.L.U., Pharmakine, and Midatech Biogune, visited NIST in October 2008 to learn more about NIST's Biosciences and Bioengineering research activities.

Workshops & Meetings

Economic Strategy for Health Care through Bio and Information Standards and Technologies
September 25, 2007
Gaithersburg, MD

This inaugural conference on Developing an Economic Strategy for Health Care through Standards and Technologies was to initiate dialogue on developing a strategic plan for the Nation to address the growing need for new technologies to help avert the impending economic crisis in Health Care and to improve quality. Understanding the vision, the technology gaps that stand in the way of achieving that vision and addressing a method for measuring market performance of technologies to fill those gaps are critical first steps in the process of planning for investments into new technologies, widely accepted standards.

12th Topical Conference on Quantitative Surface Analysis
October 12-13, 2007
Bellevue, WA

The Quantitative Surface Analysis Conference provided a forum for extended discussion on the quantitative aspects of surface analysis and on surface, interface, and thin-film characterization.

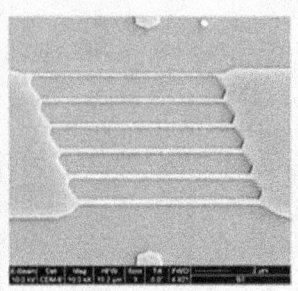

The 2nd Tri-National Workshop on Standards for Nanotechnology
February 6, 2008
Gaithersburg, MD

This workshop covered the state of nanotechnology in all three North American nations, the limits of current technology and standards – both documentary and physical. One major goal of the workshop was to increase the mutual awareness and cooperation among North American delegates to international standards bodies.

Workshop on 2D and 3D Content Representation, Analysis, and Retrieval
May 1-2, 2008
Gaithersburg, MD

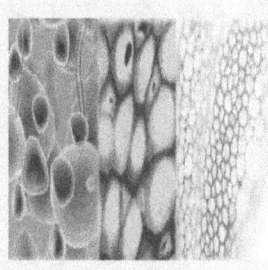

The 2nd Annual Workshop on 2D and 3D Content Representation, Analysis and Retrieval was organized with the goal of bringing the various scientific research communities together to present their recent research on Image processing, image analysis, shape analysis, indexing, data mining, metadata, ontology, interoperability tools, benchmarks and evaluation methodologies.

21st Annual Workshop on Secondary Ion Mass Spectrometry
May 13-16, 2008
San Antonio, TX

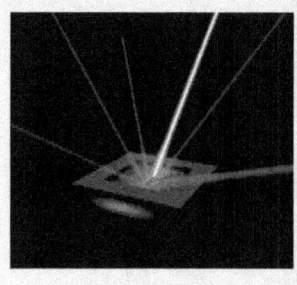

Highlights of the conference included discussions on new efforts to characterize the surface chemistry of nanoparticles and the direct molecular mapping of biological tissue samples for cancer diagnostics. Another focus area of the conference was new developments in the use of atmospheric pressure desorption electrospray ionization for direct analysis of surfaces with no sample preparation.

2008 World Metrology Day: Metrology for the Olympic Games
May 20, 2008
Gaithersburg, MD

World Metrology Day is a celebration of NIST's core foundation and purpose! This global theme for this year, "Metrology Measurement in Sport," recognized the importance of the base SI units and the need for SI traceable measurements of length, time, mass, and chemical composition with the comparison of results from sporting events over both space and time. This year's celebration of World Metrology Day Symposium at NIST focused on "Metrology for the Olympic Games," with a special emphasis on "metrology in the testing for performance enhancing drugs."

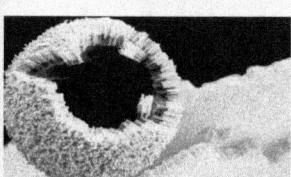

Cross Industry Issues in Nanomanufacturing Workshop
May 20-22, 2008
Gaithersburg, MD

The workshop was conducted to find common problems and common solutions specific to nanotechnology, manufacturing processes, and performance of nanomaterials in commercial products within widely different industries, including aerospace, automotive, chemical, food, forest products, medical technology, pharmaceutical, and semiconductor.

63rd Calorimetry Conference
July 3, 2008
Jersey City, NJ

The Calorimetry Conference is held annually and is a joint effort with the International Conference on Chemical Thermodynamics when the conference is held in North America. A unique aspect of the 2008 conference was the tutorials that provided an overview Isothermal Titration and Differential Scanning Calorimetry to characterize biomolecular interactions and biopolymer stability, with a focus on how microcalorimetry is being used in the discovery and development of new therapeutics.

CALCON 2008

16th Symposium on Thermophysical Properties
June 21-26, 2008
Boulder, CO

This conference is a well-established series of conferences on thermophysical properties. The Symposium is concerned with theoretical, experimental, simulation, and applied aspects of the thermophysical properties of gases, liquids, and solids, including biological systems.

Workshop on Properties Needs for Biofuels and Blends: Production, Distribution, and End Use
July 10-11, 2008
Boulder, CO

In this Workshop, invited speakers discussed property needs from the perspectives of chemical process design, engine performance, and infrastructure issues. Presented was the initial NIST work on the properties of biofuels and earlier, related work on other fuels. Breakout groups considered priorities for fuels and needed property data, and consider how best to disseminate these data to stakeholders.

Multi-agency Coordination Committee for Combustion Research Summit on Fuels
September 8-10, 2008
Gaithersburg, MD

This conference was designed to enable the future utilization of alternative fuels through research in fuel formulation, simulation, and testing methodologies. It is a direct response to concerns over the future cost, availability, and environmental impact of conventional hydrocarbon fuels.

Accelerating Innovation in 21st Century Biosciences: Identifying the Measurement Standards and Technological Challenges
October 19-22, 2008
Gaithersburg, MD

The workshop was conducted to identify and prioritize the measurement, standards and technology barriers to the realization of optimal economic and broad societal benefit from new discoveries in the Biosciences.

FINANCE & INVESTMENTS

Finance & Investments

Our Program Areas

There are ten Program Areas which the Chemical Science and Technology Laboratory (CSTL) supports. These programs expand and contract over time to meet changing needs and challenges. These programs are:

- Bioscience/Health Chemistry
- Electronics/Telecom Energy
- Environment
- International Activities
- Materials Science
- Nanotechnology
- Public Safety/Security
- Standards

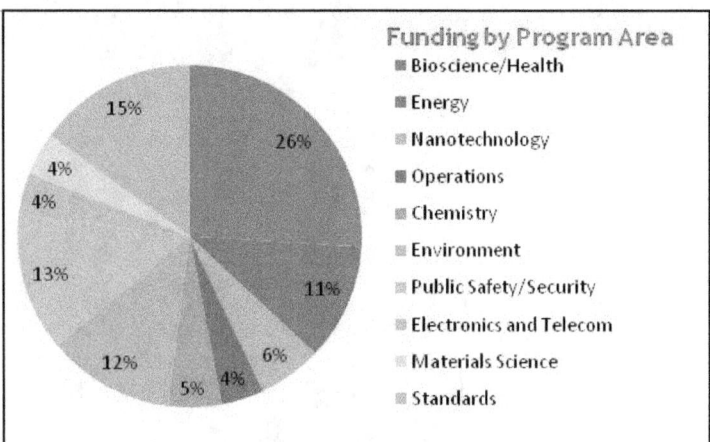

Our Strategic Investment Priority Areas

CSTL's areas of strategic emphasis are identified to capture the emerging technology in areas of importance. These are focus areas are strategically aligned with the NIST focus areas: *Bioscience/Health, Climate Change Assessment, Energy, Nanometrology*

Our Finances

CSTL operating budget for Fiscal Year 2008 was ~$90M. These funds are Congressionally Appropriated, from Other Government Agencies, and income from our Measurement Service Programs.

Our Major Customers

Customers of our measurements services, come to NIST for Standard Reference Materials (SRM), Standard Reference Data Products (SRD), and Calibration Services. Chemical standards constitute over 2/3 of ~1,400 NIST SRM types, and ~26,500 of over 32,000 NIST SRM units sold annually. 50% of SRM sales come from 5% of the individual SRMs; 50% of SRM sales are to Customers outside the U.S.; Resellers are ~10% of all sales, and are top 10 ten customers due to sales volume.

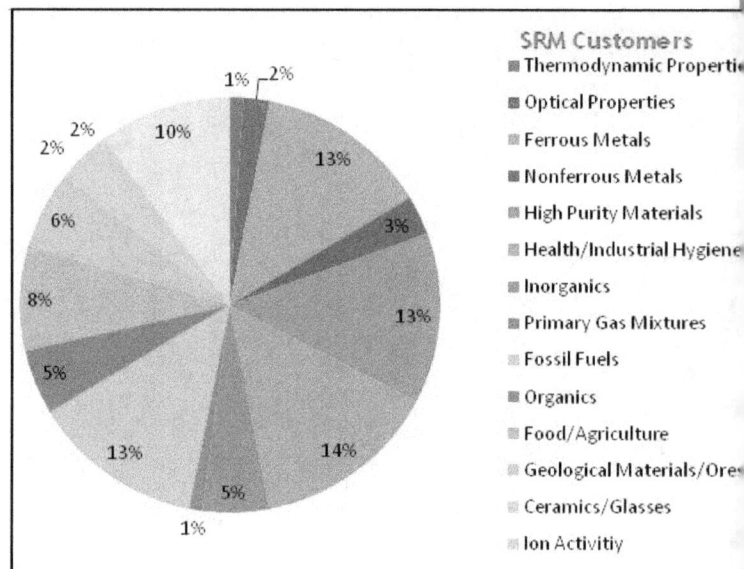

The customers of our SRD products are Instrument Manufacturers and Software Developers who incorporate NIST products into their equipment and public/commercial software packages. Our calibration customers are made up of ~150 Instrument Companies, Power Generation Companies, and U.S. Federal Government Agencies. CSTL performs ~50% of all NIST calibrations.

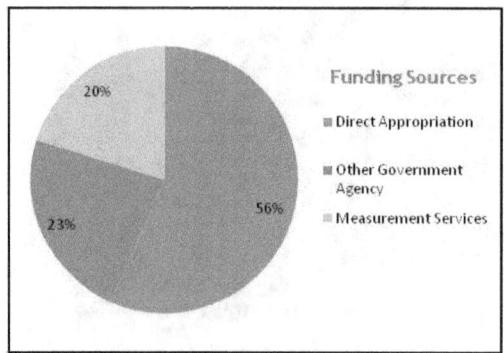

Gaithersburg, MD

HML—Charleston, SC

CARB—Rockville, MD

Boulder, CO

Our Staff
The research and services conducted in CSTL is enabled by an outstanding and diverse staff who are committed to ensuring CSTL continues as the dynamic, flexible, and innovative entity it is.

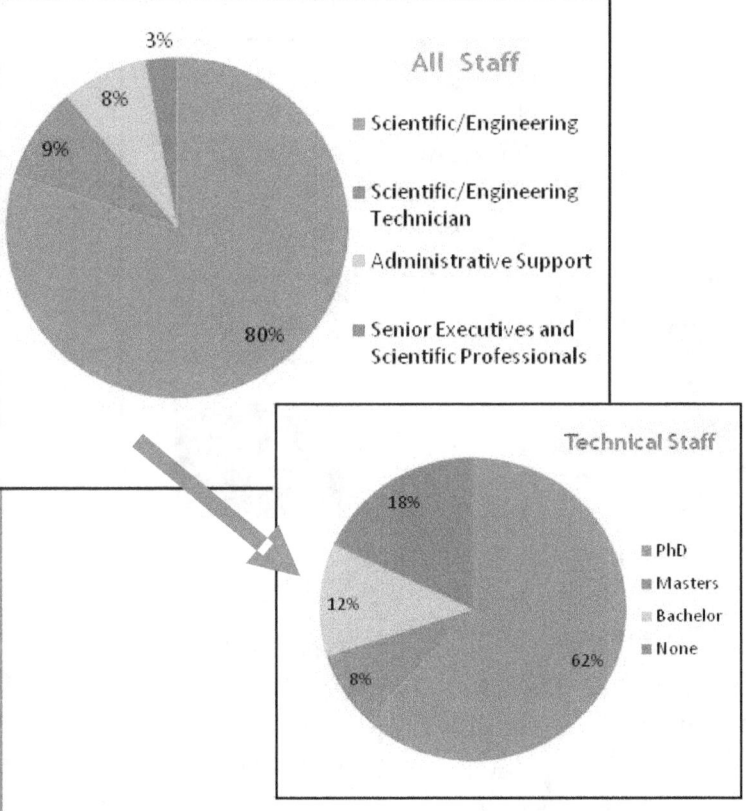

~336 Full-Time Permanent Staff
~250 Visiting Scientists

Our Research Campuses
CSTL conducts its research and services at two campuses—Gaithersburg, Maryland and Boulder, Colorado; and two joint institutes—University of Maryland Biotechnology Institute's Center for Advanced Research in Biotechnology (CARB), Rockville, Maryland; and the Hollings Marine Laboratory, Charleston, South Carolina.

Our Grants and Special Programs
Through our grants and education programs, CSTL supports science education at all levels, from undergraduate student science training and internships to postdoctoral research programs. CSTL supports ~25 undergraduate students and ~15 Postdoctoral students per year. Annual disbursements under our Grants program average $1.5M/year. Annual contributions to our Postdoctoral research programs average $3M/year.

Cooperative Research and Development Agreement (CRADA)
The CRADA is a partnering tool that allows federal labs to work with US industries, academia and organizations on cooperative R&D projects. A CRADA is the means by which NIST can provide the rights to NIST inventions conceived during the project. CSTL currently has 6 agreements with Arkema, Inc., CropLife International, Luna Innovations Inc., Network Biosystems, NYU School of Medicine, and University of Missouri.

Willie May
Director

Richard Cavanagh
Deputy Director

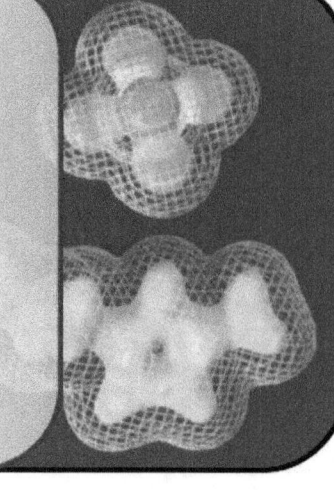

Stephen Wise
Analytical Chemistry

Laurie Locascio
Biochemical Science

Carlos Gonzalez
Chemical & Biochemical Reference Data

James Whetstone
Process Measurements

John Small
Surface & Microanalysis Science

Dan Friend
Thermophysical Properties

Authors:

May, W.E.; Cavanagh, R.R.; Poster, D.L.; and Amos, M.A.

Photo Credits:

iStockphoto.com
Microsoft Online
NIST
Shutterstock.com

Disclaimer

Certain commercial equipment, instruments, or materials are identified in this report to specify the experimental procedure adequately. Such identification is not intended to imply recommendations or endorsement by the National Institute of Standards and Technology, nor is it intended to imply that the materials or equipment identified are necessarily the best available for the purpose.

For More Information

Chemical Science and Technology Laboratory
100 Bureau Drive
MS 8300
Gaithersburg, MD 20899
301-975-8300
cstlinfo@nist.gov
www.nist.gov/cstl

October 2009

NIST
National Institute of Standards and Technology
U.S. Department of Commerce

www.ingramcontent.com/pod-product-compliance
Lightning Source LLC
Chambersburg PA
CBHW080259180526
45167CB00006B/2597